Landraces - Its Productive Conservation in Animals and Plants

Edited by Arnoldo González-Reyna and Prashant Kaushik

Published in London, United Kingdom

Landraces - Its Productive Conservation in Animals and Plants
http://dx.doi.org/10.5772/intechopen.100873
Edited by Arnoldo González-Reyna and Prashant Kaushik

Contributors
Paul Kweku Tandoh, Irene Akua Idun, Bridget Yayra Bemanu, Nkanyiso J. Sithole, Khayelihle Ncama, Lembe S. Magwaza, Nozipho M. Motsa, Jacob Mashilo, Ayanda Maqhashu, Rimbilana Shingange, Fhulufhelo Vincent Ramukhithi, M.C. Samuel Tabe Roldan, Marco Antonio Esteves Luna, Leonel Avendaño Reyes, Arnoldo González-Reyna, Fredy Minor Pérez, Francisco Javier Trejo Meza, Julio Cesar Camacho Ronquillo, Sulukhan Temirbekova, Ivan Kulikov, Yuliya Afanaseva, Elena Kalashnikova, Natalia Ionova, Anand Kumar, Prashant Kaushik

First published in London, United Kingdom, 2023 by IntechOpen
IntechOpen is the global imprint of INTECHOPEN LIMITED, registered in England and Wales, registration number: 11086078, 5 Princes Gate Court, London, SW7 2QJ, United Kingdom

British Library Cataloguing-in-Publication Data
A catalogue record for this book is available from the British Library

Additional hard and PDF copies can be obtained from orders@intechopen.com

Landraces - Its Productive Conservation in Animals and Plants
Edited by Arnoldo González-Reyna and Prashant Kaushik
p. cm.
Print ISBN 978-1-83768-158-7
Online ISBN 978-1-83768-159-4
eBook (PDF) ISBN 978-1-83768-160-0

Meet the editors

Arnoldo Gonzalez Reyna has a bachelor's degree in Agronomy from the Universidad Autónoma de Tamaulipas (UAT), Mexico, an MSc from Utah State University, USA, and a Ph.D. from the University of Saskatchewan, Canada, where he also completed a postdoctoral fellowship. He is Professor Emeritus at UAT, with an academic and research career spanning more than 37 years. Dr. González Reyna worked with Criollo cattle and hair sheep for the Mexican Association of Animal Production for 13 years. He has authored numerous publications, including indexed and refereed articles, book chapters, and books. His areas of expertise are ruminant physiology and endocrinology of reproduction, animal production systems, and biotechnology of reproduction. He continues teaching and counseling graduate students at UAT and working with private producers in Northeast México.

Dr. Prashant Kaushik holds adjunct faculty positions at three universities and is a genome data engineer for a private company. He holds a BS (Hons) in Agriculture, a PG diploma in Floriculture and Landscaping, an MS in Vegetable Science, and an MBA in Production and Operations Management. After obtaining a Ph.D. in Biotechnology from Universitat Politecnica de Valencia, Spain, Dr. Kaushik was a bioinformatician at Nagano University and Yokohama Ueki, Japan. He has published 150 research articles and an equivalent number of non-indexed research publications. He has written six books and fourteen book chapters. Among his research interests are molecular biology, plant breeding, genomics, landraces, and population genetics.

Contents

Preface

My first encounter with landraces was with a Criollo (or Creole) cattle (developed in Central America, called Criollo Lechero Centroamericano) and hair sheep (in México, known as Pelibuey and Blackbelly). It was not until I began editing the first chapter of this book that I realized that I had been living with animal landraces all my life. In later years, and by trade, I learned and worked with some plant landraces (corn, sorghum, sunflower, and some forages and grasses) to complete my agronomy degree.

Landraces are local species of plants and animals that have developed and adapted over time to a localized geographic region. They are important components of our environment and play an important role in the sustainability of the environment.

Every continent on Earth is composed of different regions and every region is composed of several eco-regions or ecosystems. As such, this book discusses landraces in the context of different regions and countries and makes no attempt to generalize or impose conclusions for global application.

This book is organized into two sections on animal landraces and plant landraces. Section 1 includes two chapters that describe the use of ruminants in tropical regions. Section 2 includes four chapters on the benefits of the use of indigenous plants and strategies for improving wheat production.

I would like to thank the contributing authors for their efforts in writing the chapters. I also acknowledge and thank the support provided by the academic and editorial team, as well as the trust and responsibility granted to me as editor by IntechOpen.

Arnoldo González-Reyna
Autonomous University of Tamaulipas,
Cd. Victoria, Tamaulipas, México

Prashant Kaushik
Institute of Conservation and Improvement of Valencian Agrodiversity,
Polytechnic University of Valencia,
Valencia, Spain

Section 1

Animal Production Management

Chapter 1

The Use of Hormonal Assay, Phenotypic Morphometry, and CASA Semen Analysis to Estimate Attainment of Puberty in Indigenous Ram Lambs

Rimbilana Shingange, Fhulufhelo Vincent Ramukhithi and Ayanda Maqhashu

Abstract

Ruminant landraces that are locally adapted have become crucial for sustainable farming considering climate change. This study sought to evaluate the commencement of reproductive capability of Bapedi, Namaqua-Afrikaner and Zulu ram lambs. Data were collected from a total of 21 ram lambs (7/breed) starting from 3–8 months of age. From four months of age, the scrotal circumference of rams was measured using a flexible tape and it was found that it differed significantly between breeds ($P < 0.05$). Blood was collected from the jugular vein using a 21-gauge needle and red cap vacutainers. Blood serum testosterone levels were obtained using a competitive enzyme immunoassay. It was found that Bapedi, Namaqua-Afrikaner and Zulu ram lambs had similar blood serum testosterone levels ($P > 0.05$). At 7 months, semen was collected using an electro-ejaculator and analysed using a Sperm Class Analyser©. There were significant differences found in semen quality between the studied breeds for various semen parameters ($P < 0.05$). There was a weak positive correlation between blood serum testosterone and scrotal circumference ($r = 0.30$). Conclusively, this study highlights the importance of characterisation for the conversation of landrace breeds.

Keywords: blood serum testosterone, scrotal circumference, semen quality, male reproduction, hormonal assay, indigenous ram lambs

1. Introduction

Male animals contribute 60% to 70% of a flock's genetics [1]; this is because in many smallholder farming systems, a male animal services more than one female during breeding at a ratio of 1 ram: 45 ewes [2]. This outlines the importance of having 'productive' rams in a flock: 'Productive' refers to the animal's overall health and is a prerequisite for reproductive efficiency [1]. Having reproductive male is important even when assisted reproductive technologies (ARTs) like Artificial insemination (AI) are used; even in small flocks like those of smallholder farmers.

In the Southern African Development Community (SADC), indigenous sheep play an integral role in the income and sustenance of smallholder farmers and livestock keepers [3]. This is partially due to the arid/semi-arid environment of sub-Saharan Africa - indigenous breeds like the Bapedi, Namaqua-Afrikaner and Zulu are particularly suited to South African environmental conditions due to their continued adaptation. However, despite this adaptation, indigenous breeds are continuously being replaced by exotic breeds like the Merino due to better production [4]. This problem is exacerbated by there being no or slow genetic progress for the indigenous genetic resources used by livestock keepers and smallholder farmers [3].

South Africa houses 46 of the 109 different sheep breeds in sub-Saharan Africa [5]; it also houses 19.9 million of the 39 million sheep in the SADC region [6]. Three of South Africa's oldest sheep breeds are the Bapedi, Namaqua-Afrikaner and Zulu sheep: The Bapedi sheep have small frames, they are polled, and have fat-tails along with long legs and a shallow body; they are hardy and disease tolerant [7]. Similar in general phenotype but multicoloured, the Namaqua-Afrikaner is South Africa's oldest indigenous sheep breed [8, 9], and is used primarily in smallholder farming systems [10] and has a fat tail. They are adapted to the South African Karoo's environmental conditions [4]. Namaqua-Afrikaners are reported to be endangered: the number of breeding females ranges from a hundred to one thousand and the number of breeding males ranges between six and twenty [11]. Thus, urgent intervention to conserve this breed is necessary. Lastly, the Nguni sheep of Zululand, Zulu sheep, are generally dark coloured (sometimes with spots), with fat tails and long legs and shallow bodies. They are classified as insecure [12] because of the limited available information on their characteristics. The Zulu sheep are, however, heat, humidity, and tick tolerant [12].

Indigenous breeds are very important, and their genetic erosion is problematic because of the loss of their adaptability [3]. With climate change an ever-growing concern in sub-Saharan Africa, it is thus important to maintain breeds that are adapted.

Testosterone is considered the most fundamental hormone in male reproduction because although gonadotrophin-releasing hormone (GnRH) and luteinising hormone (LH) play important roles (for example, GnRH acts on the anterior pituitary and LH on the testes), testosterone is important for sexual behaviour [13], secondary sex characteristics and sperm production [14]. Testosterone levels can be ascertained from blood serum or from testicular tissue [13].

Morphologically, scrotal circumference size has been reported to be positively correlated with sperm production as the larger the scrotal circumference, the larger the surface area of testes available for spermatogenesis (testes are made up of 80% seminiferous tubules) [13]. Spermatogenesis is the process by which germ cells in the seminiferous tubules form haploid spermatozoa [15]. It is important to note that testosterone is produced by the Leydig cells of the testes under stimulus from LH, which increase in number the larger the testes are [13]. Similarly, follicle-stimulating hormone (FSH) acts on the testes' spermatogonia to initiate and support spermatogenesis, and on the seminiferous tubules' sertoli cells that assist in the production of sperm [1] - the larger the testes, the larger the available surface area for these processes [13].

This study aimed to use hormonal assays, phenotypic morphometry, and CASA semen analysis to estimate the attainment of puberty in indigenous ram lambs as the age at which a ram attains puberty may be crucial for efficient animal production. This is because the ram lamb can now contribute to the productivity of the herd; A herd which, if a landrace breed, may achieve superior productivity parameters compared to exotic breeds due to their adaptability and hardiness to their respective environments.

2. Materials and methods

Live animal work was conducted at the Agricultural Research Council Animal Production Institute, in Irene, South Africa and laboratory work at the University of Pretoria, in Pretoria, South Africa.

2.1 Statistical analysis

Collected numerical data were analysed using Statistical Analysis Software (SAS University Edition, SAS Institute Inc., 2020). N-way Analysis of Variance (ANOVA) tests was performed with Tukey-Kramer adjustments for multiple comparisons, using PROC GLM with the STDERR statement, to obtain the least square means ± standard error values. Data were graphed using Microsoft Excel 2019. In addition, Pearson correlations were also obtained using Statistical Analysis Software (SAS University Edition, SAS Institute Inc., 2020).

2.2 Ethical approval

Ethical approval was obtained from the University of Pretoria: NAS207/2020, as well as from the Agricultural Research Council: APAEC 2020/07. The study also received Section 20 approval from the South African Department of Agriculture, Forestry and Fisheries.

2.3 Experimental animals

Seven ram lambs of each studied breed (Bapedi, Namaqua-Afrikaner and Zulu) were used in this study, which commenced at 4 months of age (for blood serum testosterone and scrotal circumference) and at 7 months of age (for semen quality) until 8 months of age. The twenty-one ram lambs were kept in a semi-extensive habitat with *ad-libitum* access to water in metal troughs and grazing on natural pasture.

2.3.1 Scrotal circumference

Scrotal circumference was measured using a flexible measuring tape while the ram was in a standing position in a crush. The technician pulled the testicles ventrally into the scrotum and measured the circumference of the scrotum at it its widest; the measuring tape was snug on the scrotum. These measurements were done bi-weekly.

2.3.2 Blood sampling

Blood samples were also collected bi-weekly from the jugular vein: A 21-gauge syringe and red-cap vacutainers were used for all ram lambs, and blood serum was harvested by pipetting the collected blood 24 hours after collection and stored at −20°C until analysis.

2.3.3 Semen collection and analysis

An electro ejaculator was used to collect semen from ram lambs which consisted of a variable source of electric current which varied from 0 to 5 volts and 0.5 to 1.0 amps at 7 months of age. After hair from around the sheath was shaved and the prepuce was cleaned with a sterile paper towel to prevent contamination while the ram was placed in a sitting position, the ram was placed on its side and the electrode was also cleaned with a sterile paper towel. The electrode was then lubricated and inserted transrectally and positioned over the accessory glands.

Semen was collected using 15 mL graduated tubes that were pre-warmed to 37 degrees Celsius and then transported within 30 minutes to the Agricultural Research Council's mobile laboratory for analysis: To determine semen volume, measurements on the graduated tube were read before dilution in mL; to determine semen concentration, a spectrophotometer was used (Jenway, United Kingdom). A square cuvette was filled with 3 mL of sodium citrate solution and placed in the spectrophotometer for at least 30 seconds. Raw semen (20 μL) was added in a square cuvette containing the sodium citrate solution, and again placed in the spectrophotometer to read the absorbance. The absorbance was used to determine the final sperm cell concentration with the aid of a formula: 151 × (25.97 X absorbance − 0.3) where 151 is the dilution factor - the final sperm cell concentration was recorded in millions per millilitre; to determine membrane integrity of sperm cells, hyperosmotic osmotic swelling test (HOS) was performed wherein the samples were evaluated under a phase contrast microscope (400x) and 200 sperm cells per slide were counted. Sperm cells with swollen and coiled tails were considered intact (Naing *et al.*, 2010); to determine sperm cell abnormalities and viability, 7 μL of semen was stained with 20 μL of nigrosine-eosin stain in an Eppendorf tube and 5 μL of that mixture was smeared onto a microscopic slide.

3. Results

3.1 Scrotal circumference

The scrotal circumferences of South African indigenous ram lambs are reported in **Figure 1**: There existed a significant difference between the scrotal circumference of Zulu ram lambs at 4 months of age and 7.5 and 8 months of age (15.7 ± 1.0, 20.7 ± 1.0 and 20.7 ± 1.0, respectively) ($P < 0.05$), compared to Bapedi and Namaqua-Afrikaner ram lambs which had no significant within-breed differences ($P > 0.05$). However, there also existed significant differences in between-breed differences ($P < 0.05$): Bapedi and Zulu ram lambs had SC significantly higher than Namaqua-Afrikaner ram lambs at 4 months of age and again at 5 to 8 months of age. In addition, the SC of Bapedi ram lambs ranged from 19.8 ± 1.7 cm to 23.5 ± 1.8 cm; that of

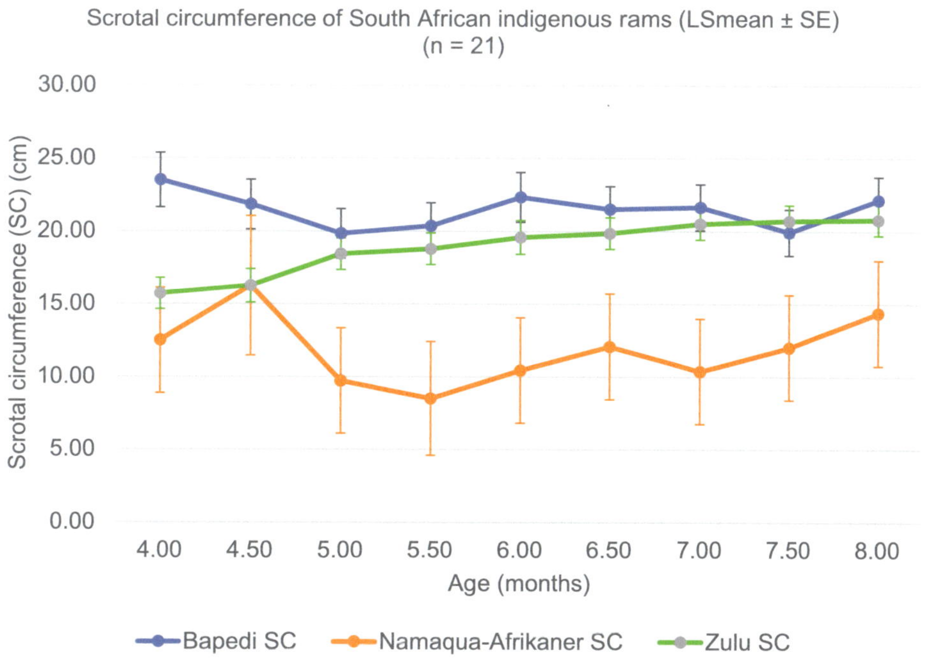

Figure 1.
Scrotal circumference of South African indigenous rams.

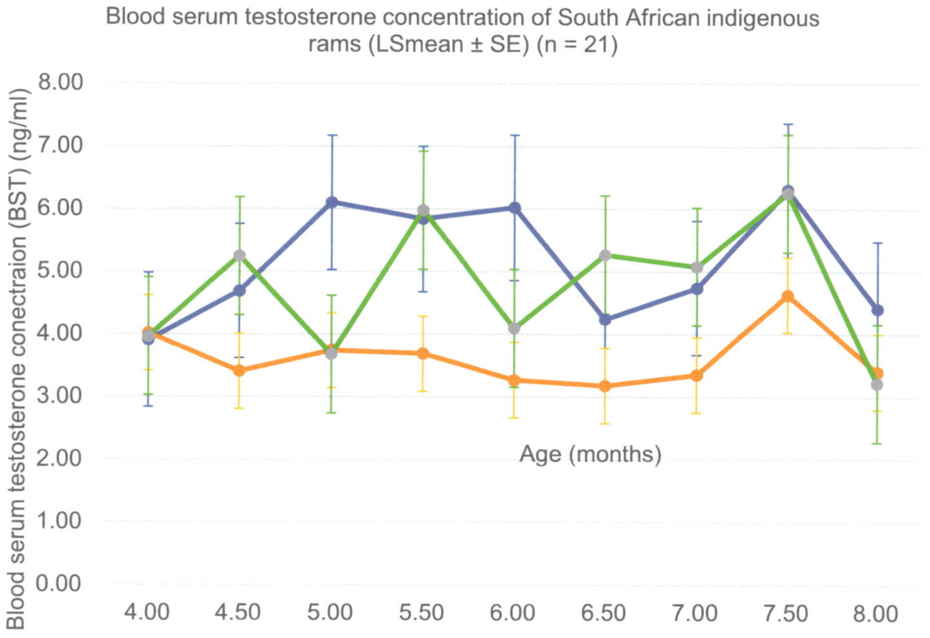

Figure 2.
Blood serum testosterone of South African indigenous rams.

Namaqua-Afrikaner ram lambs from 8.5 ± 3.9 cm to 16.2 ± 4.7 cm; and that of Zulu ram lambs from 15.7 ± 1.0 cm to 20.7 ± 1.0 cm.

3.2 Blood serum testosterone

The blood serum testosterone concentrations of South African indigenous ram lambs are reported in **Figure 2**: There was a significant difference between the blood serum testosterone concentration of Bapedi and Namaqua-Afrikaner ram lambs at 6 months of age (6.0 ± 1.1 and 3.2 ± 0.6 ng/ml, respectively). The blood serum testosterone ranges of Bapedi sheep were 3.9 ± 1.0 ng/ml to 6.3 ± 1.0 ng/ml; that of Namaqua-Afrikaner sheep were 3.1 ± 0.6 ng/ml to 4.6 ± 0.6 ng/ml, and that of Zulu sheep were 3.2 ± 0.9 ng/ml to 6.2 ± 0.9 ng/ml.

3.3 Semen parameters

The semen characteristics of Namaqua-Afrikaner, Bapedi and Zulu ram lambs are compared in **Table 1**: At ages 7, 7.5 and 8 months of age, there was no significant difference between the semen volumes of any of the three breeds ($P > 0.05$). In addition, the semen volumes ranged from 0.3 ± 0.06 mL to 0.6 ± 0.1 mL. Contrastingly, there was a significant difference ($P < 0.05$) between the sperm cell concentrations of Bapedi sheep at 7 and 8 months of age (1.4 ± 0.6 x109/mL and 4.3 ± 0.5 x109/mL, respectively); and of Zulu ram lambs between 7 and 8 months (1.5 ± 0.5 x109/mL and 5.4 ± 0.5 x109/mL, respectively), and between 7.5 and 8 months (2.3 ± 0.5 x109/mL

Breed		Bapedi			Namaqua-Afrikaner			Zulu		
Age (months)		7	7.5	8	7	7.5	8	7	7.5	8
Semen volume (mL)		0.6 ± 0.1	0.6 ± 0.1	0.5 ± 0.1	0.6 ± 0.1	0.5 ± 0.1	0.4 ± 0.1	0.3 ± 0.06	0.4 ± 0.06	0.4 ± 0.06
Sperm cell concentration ($\times10^9$/mL)		1.4 ± 0.6^a	2.3 ± 0.5	4.3 ± 0.5^b	1.1 ± 1.6	2.4 ± 1.6	4.9 ± 1.6	1.5 ± 0.3^a	2.3 ± 0.3^a	5.4 ± 0.3^b
Membrane integrity (%)	Intact	78.8 ± 4.2	81.6 ± 3.5	84.0 ± 3.5	81.0 ± 4.0	80.7 ± 4.0	84.5 ± 4.0	79.5 ± 5.0	85.1 ± 5.4	86.4 ± 5.0
	Non-intact	21.1 ± 4.1	18.3 ± 3.4	15.8 ± 3.4	19.0 ± 1.8	16.7 ± 1.8	15.5 ± 1.8	21.3 ± 4.7	14.9 ± 5.1	13.5 ± 4.7
Viability (%)	Live	79.0 ± 5.3	82.9 ± 4.3	85.3 ± 4.3	76.5 ± 11.4	80.5 ± 11.4	83.7 ± 11.4	82.8 ± 7.7	77.7 ± 8.5	86.0 ± 7.7
	Dead	21.0 ± 5.3	17.0 ± 4.3	14.6 ± 4.3	23.5 ± 11.4	19.5 ± 11.4	16.2 ± 11.4	17.1 ± 7.7	22.3 ± 8.5	14.0 ± 7.7
Abnormalities (n)	Head	1.2 ± 0.2^a	0.5 ± 0.2	0.3 ± 0.2^b	2.0 ± 1.3	1.0 ± 1.3	0.5 ± 1.3	1.3 ± 0.5	0.8 ± 0.5	0.1 ± 0.5
	Tail	4.2 ± 1.1	1.1 ± 0.9	1.5 ± 0.9	4.0 ± 2.2	4.5 ± 2.2	2.5 ± 2.2	4.1 ± 0.7^a	1.4 ± 0.8	1.0 ± 0.7^b
	Mid-piece	0.0 ± 0.1	0.0 ± 0.1	0.3 ± 0.1	0.0 ± 0.2	0.0 ± 0.2	0.5 ± 0.2	0.0 ± 0.0	0.0 ± 0.0	0.0 ± 0.0

[a,b]*Values with different superscripts within the same cell in the same row differ significantly ($P < 0.05$).*

Table 1.
Semen parameters of South African indigenous rams.

Parameter		Blood serum testosterone	Scrotal circumference
Blood serum testosterone		1.00	+0.30
Scrotal circumference		+0.30	1.00
Semen volume		+0.05	+0.33
Sperm cell concentration		−0.40	+0.07
Membrane integrity	Intact	−0.06	+0.05
	Non-intact	+0.03	−0.05
Viability	Live	−0.02	+0.13
	Dead	+0.02	−0.13
Abnormalities	Head	+0.35	−0.35
	Tail	−0.0011	+0.15
	Midpiece	+0.15	+0.05

Table 2.
Pearson correlation coefficients for blood serum testosterone and scrotal circumference.

and 5.4 ± 0.5 x109/mL, respectively). In addition, all three breeds had similar percentages of intact and non-intact membranes under membrane integrity ($P > 0.05$). This is the same for live and dead percentages under viability. Lastly, in terms of abnormalities: There was a significant difference ($P < 0.05$) between the number of head abnormalities of Bapedi sheep at 7 and 8 months (1.2 ± 0.5 and 0.3 ± 0.4). Similarly, for Zulu ram lambs, there was a significant difference between the tail abnormalities at 7 and 8 months of age (4.1 ± 0.7 and 1.0 ± 0.7) ($P < 0.05$).

3.4 Pearson correlations for blood serum testosterone and scrotal circumference

Pearson correlation coefficients for blood serum testosterone, scrotal circumference, semen volume, semen concentration, membrane integrity, viability and abnormalities are shown in **Table 2**: There exists a weak positive correlation of r = 0.30 between blood serum testosterone and scrotal circumference. There also exists varying positive correlations between blood serum testosterone and semen volume, non-intact membrane integrity, dead viability, head abnormalities and midpiece abnormalities (0.05, 0.03, 0.02, 0.35 and 0.15 respectively); and between scrotal circumference and semen volume, semen concentration, intact membrane integrity, live viability, tail abnormalities, and midpiece abnormalities (0.33, 0.07, 0.05, 0.13, 0.15, and 0.05, respectively).

4. Discussion

4.1 Scrotal circumference

Bapedi, Namaqua-Afrikaner and Zulu ram lambs had statistically significant differences between breeds for scrotal circumferences measurements ($P < 0.05$) for ages 4 to 8 months. This heterogeneity between breeds was unexpected, however, the cause of this heterogeneity may be due to individual animal variation [16]. All three studied breeds had different scrotal circumferences from 4 to

8 months ($P < 0.05$). This may be because testosterone is responsible for sexual maturity and interest and is produced by the Leydig cells of the testes. Leydig cell parameters (like the number of Leydig cells per gram of testis, the total number of Leydig cells per testis and the per cent cell volume of Leydig cell nuclei) are correlated significantly with testosterone levels [17] and testosterone is responsible for secondary sex characteristics like sexual behaviour [18]. Thus, because the Bapedi, Namaqua-Afrikaner and Zulu breeds have dissimilar scrotal circumferences, they will have correspondingly dissimilar Leydig cells and subsequent testosterone production. This suggests that these breeds would attain puberty at different ages.

Söderquist and Hultén [19] found that ram lambs of the Gotlandic breed, a Swedish breed, had a scrotal circumference of 28.9 ± 1.9 cm, which was higher than the scrotal circumferences found in this study. Despite the differences in scrotal circumference between South African indigenous sheep and Icelandic sheep, scrotal circumference is still a useful selection indicator: Males with larger testes are likely to sire daughters that reach puberty at an early age and ovulate more ova during oestrus [19]. This would be especially useful for smallholder farmers and livestock keepers as they would be able to get more lambs from one ewe, both because the age at first lambing is lower and the likelihood of multiple births is higher. However, the significant difference in the scrotal circumferences of Zulu ram lambs at 4 months of age and 7.5 and 8 months of age ($P < 0.05$) indicates that there is a substantial change in SC for Zulu ram lambs between these ages and may act as a selection indicator.

Notably, the significant differences between breeds found from 4 to 8 months have a distinct trend: Bapedi ram lambs generally had the highest SC (except at 7.5 months), and Namaqua-Afrikaner ram lambs had the lowest SC (at all months). This difference in SC may be due to the Namaqua-Afrikaner's late sexual maturity [20]. Late-maturing breeds are physiologically younger at the same chronological age and thus would weigh the same as Bapedi and Zulu sheep but be at a different physiological age with regards to reproduction.

In addition, SC is affected by the season of the year, breed and body condition but would usually be at a maximum peak during the breeding season [21]. This study was conducted between October and the following June in South Africa – meaning Summer through to Autumn and early Winter. Because Autumn and Spring are the breeding seasons of sheep, and an increase (albeit insignificant ($P > 0.05$) in scrotal circumference was observed over this time span, it is possible that there is a correlation between breeding season and scrotal circumference.

4.2 Blood serum testosterone concentration

Bapedi, Namaqua-Afrikaner and Zulu ram lambs had similar blood serum testosterone levels at all ages ($P > 0.05$) except at 6 months of age. This means that although blood serum testosterone levels fluctuated for each breed as they aged from 4 to 8 months of age, this fluctuation was not one that resulted in a significant difference between most samples, both within and between breeds. Thus, the three studied breeds displayed homogeneity in blood serum testosterone levels i.e., there were no significant within-breed differences in BST concentration and minimal between-breed differences ($P < 0.05$). In addition, testosterone is needed to initiate spermatogenesis at puberty and to maintain this process in the adult [17], thus the presence of blood serum testosterone in all three breeds may indicate the attainment of puberty. Strikingly, Bapedi, Namaqua-Afrikaner and Zulu ram lambs all had the highest sampled blood serum testosterone levels at 7.5 months: Spermatogenesis and its sustenance take 2 months [22], thus this spike in blood serum testosterone may indicate the attainment of puberty [18] already at 5.5 months.

4.3 Semen parameters

Bapedi, Namaqua-Afrikaner and Zulu ram lambs had statistically nonsignificant differences in semen volumes from each other from the first semen analysis at 7 months to the last semen analysis at 8 months of age ($P > 0.05$); Homogeneity among the breeds was shown. It is also important to note that the semen volume observed in this study was expected to be low as ram lambs have lower semen volumes compared to rams due to the parabolic nature of semen volume output in rams i.e., semen volume, while the ram is very young and very old, is low [23]. Also, Rege *et al.* [24] found the semen volumes of Ethiopian highland sheep at 6 months to have a semen volume of 0.30 ± 0.67 mL, which corresponds with the semen volume range found in this study at 7 months of age (0.3 ± 0.06 mL to 0.6 ± 0.1 mL).

Contrastingly, Bapedi sheep had a significant difference in sperm cell concentration at 7 and 8 months of age and Zulu sheep at 7 and 8 months of age, and at 7.5 and 8 months of age ($P < 0.05$), with the sperm cell concentration having increased at 8 months. This means that there was still not a significant difference between breeds, but rather within a breed; Bapedi and Zulu ram lambs' sperm cell concentrations thus increased from 7 months to 8 months, as was to be expected if it is considered that semen characteristics improve with age [24]. In addition, at 8 months of age, both Bapedi and Zulu ram lambs showed a peak in sperm cell concentrations. This peak can be presumably attributed to other factors that occurred 2 months prior to the ram lambs being 8 months old. This is because the testicular function is slow in its development, and because spermatogenesis takes 2 months to complete from the creation of spermatogonia to the creation of spermatozoa [22]. It is interesting to note, however, that this change/spike did not occur for Namaqua-Afrikaner ram lambs: It was observed that Namaqua-Afrikaners are late maturing and would then not have had the prerequisite trigger that Bapedi and Zulu ram lambs had at 6 months of age that resulted in increased semen concentration at 8 months of age; secondly, only an average of 29% of Namaqua-Afrikaner ram lambs produced semen during semen collection during the study period, compared to an average of 57% of Bapedi and 71% of Zulu ram lambs per collection day; thirdly, in this study, the underdevelopment of many Namaqua-Afrikaner ram lambs' glans penises was observed. This anatomical anomaly meant that many Namaqua-Afrikaner ram lambs could not be collected; fourthly, various Namaqua-Afrikaner ram lambs in this study had no testes i.e., scrotal circumference could not be measured because the scrotum felt like an empty sack when palpated. This may be because the testes have not yet descended because of the late sexual maturity observed for Namaqua-Afrikaners.

Bapedi, Namaqua-Afrikaner and Zulu ram lambs had a statistically nonsignificant difference in the percentage of intact and non-intact membranes as well as live and dead percentages ($P > 0.05$). This further supports the homogeneity of these breeds and suggests that all three breeds will have the same fertilizing ability [25]. Rege *et al.* [24] found that Ethiopian highland sheep at 6 months of age produced semen with 29.4 ± 6.0 % Dead percentage; this was higher than that of South African indigenous sheep at 7 months (17.1 ± 7.7 % to 23.5 ± 11.4 %). Moreover, because the semen morphology of all three breeds is above 50% but below 90% [26], the semen morphology of all three breeds is satisfactory but not exceptional from the first CASA analysis at 7 months of age, further suggesting that Bapedi, Namaqua-Afrikaner and Zulu ram lambs have attained puberty.

With regards to abnormalities, Bapedi ram lambs had a significant decrease in the number of head abnormalities at 7 and 8 months and Zulu ram lambs had a significant decrease in the number of tail abnormalities at 7 and 8 months of age ($P < 0.05$). With Namaqua-Afrikaner ram lambs having a higher number of abnormalities, it is possible that as with sexual behaviour activities, Namaqua-Afrikaner ram lambs are physiologically younger and thus have semen with more abnormalities [27]. Colas [27] found that ejaculates from younger ram lambs have a greater number of abnormal sperm cells in terms of morphology due to incomplete spermatogenic activity and incomplete epidydimal maturation.

4.4 Pearson correlations for blood serum testosterone concentration and scrotal circumference

As shown in **Table 2**, the strongest positive correlation is that between scrotal circumference and blood serum testosterone ($r = 0.30$). This means that as scrotal circumference increases, so does blood serum testosterone concentration. This corroborates the claim that Bapedi and Zulu ram lambs have significantly ($P < 0.05$) higher blood serum testosterone because of their larger scrotal circumferences. Saaed and Zaid [13] corroborate the findings of this study by stating that 'the testis of sheep release testosterone that elevates with increasing testicular weight and age until puberty and maturity age.'

For the Pearson correlations of semen characteristics, there exists a positive correlation between blood serum testosterone and scrotal circumference. It can then be assumed that ram lambs with higher semen volume will have higher blood serum testosterone and scrotal circumference as well ($r = 0.05$ and $r = 0.33$, respectively). The positive correlation between scrotal circumference and semen volume (0.33) is corroborated by Dombo [28] who reported that due to a larger site for semen production and storage, small testes produce a smaller volume of semen than big testes. In addition, there exists correlations between semen concentration and blood serum testosterone ($r = -0.40$) and SC ($r = 0.07$). This means that as semen concentration increases, blood serum testosterone decreases while scrotal circumference increases.

There is also a positive correlation between coiled tails (intact membranes) and scrotal circumference ($r = 0.05$), but a very weak negative correlation with blood serum testosterone ($r = -0.06$). This may mean that better quality semen (that with intact membranes) comes from ram lambs with higher scrotal circumferences; however, there exists an antagonism between blood serum testosterone and intact membranes. Similarly, there exists a negative correlation between straight tails (non-intact membranes) and SC (−0.05), and a very weak positive correlation with blood serum testosterone (0.03). This further supports the claim that there may exist an antagonism between blood serum testosterone and hyperosmotic swelling.

Interestingly, there exists a negative correlation between live viability percentage and blood serum testosterone (−0.02): This also supports the claim that higher blood serum testosterone may be antagonistic to positive semen characteristics; this is also supported by the very weak positive correlation between dead viability percentage and blood serum testosterone (0.03). However, there does exist very weak positive correlations between live spermatozoa percentage and scrotal circumference (0.13), and correspondingly, there exist negative correlations between dead spermatozoa percentage and scrotal circumference (−0.13). This suggests that ram lambs with higher scrotal circumferences will present a larger proportion of live semen, as in Bapedi and Zulu ram lambs. In terms of head abnormalities, there is a weak negative correlation with scrotal circumference (−0.35) and a weak positive correlation with blood serum testosterone (0.35). This means that the higher the ram lamb's BW and SC, the lower their head abnormalities, and the higher their blood serum testosterone, the higher their head abnormalities. These correlations are corroborated by the r of tail abnormalities: There is a very weak positive correlation between tail abnormalities and scrotal circumference (0.15), and a very weak negative correlation with blood serum testosterone (−0.0011), meaning that the higher the scrotal circumference of a ram lamb, the higher the tail abnormalities; and the higher the blood serum testosterone, the fewer tail abnormalities. The relationships between the tail and head abnormalities are inverse. This is also true for r of midpiece abnormalities: There are very weak positive correlations between midpiece abnormalities and blood serum testosterone (0.15) and scrotal circumference (0.05). This means that as blood serum testosterone and scrotal circumference increase, so do midpiece abnormalities.

5. Conclusions

Bapedi, Namaqua-Afrikaner and Zulu sheep had different scrotal circumferences but similar blood serum testosterone levels. This lack of significant difference is primarily, a testament to their homogeneity by virtue of age, and secondarily, by virtue of all being indigenous Southern African breeds. In addition, all the studied breeds produced satisfactory semen at 7 months of age. There existed a positive Pearson correlation between blood serum testosterone and scrotal circumference (r = 0.30) and there existed a correlation between all compared parameters. To increase the much-needed conservation and characterisation of these breeds, it is recommended that further studies be conducted on them for robust conservation strategies.

Acknowledgements

Thank you to the Agricultural Research Council Animal Production Institute, Irene, for use of their animals, and their students and laborers for any help received and for use of their facilities.

Thank you to Agriseta, NRF-Thutuka and the Department of Agriculture, Land Reform and Rural Development (DALRRD) for financial support.

Declaration of interest

None

Author details

Rimbilana Shingange[1], Fhulufhelo Vincent Ramukhithi[2] and Ayanda Maqhashu[3*]

1 Department of Animal Sciences, University of Pretoria, South Africa

2 Agricultural Research Council, Irene, South Africa

3 Department of Animal Science, University of the Free State, Bloemfontein, South Africa

*Address all correspondence to: maqhashua@ufs.ac.za

References

[1] Edmondson MA, Roberts JF, Baird AN, Bychawski S, Pugh DG. Theriogenology of sheep and goats. In: Sheep and Goat Medicine. Missouri, USA: Elsevier; 2012. pp. 150-230

[2] Gizaw S, van Arendonk JAM, Valle-Zárate A, Haile A, Rischkowsky B, Dessie T, et al. Breeding programmes for smallholder sheep farming systems: II. Optimization of cooperative village breeding schemes. Journal of Animal Breeding and Genetics. 2014;**131**(5):350-357

[3] Molotsi AH, Dube B, Cloete SWP. The current status of indigenous ovine genetic resources in southern Africa and future sustainable utilisation to improve livelihoods. Diversity. 2019;**12**(1):14-20

[4] Snyman MA, Herselman MJ. Comparison of productive and reproductive efficiency of Afrino, Dorper and Merino sheep in the False Upper Karoo. South African Journal of Animal Science. 2005;**35**(2):98-108

[5] African Union – Interafrican Bureau for Animal Resource. Strategic Plan 2010-2014. Nairobi, Kenya: African Union; 2009

[6] Department of Agriculture, Forestry and Fisheries. Abstract of Agricultural Statistics. Pretoria, South Africa: Department of Agriculture, Forestry and Fisheries; 2019

[7] Maqhashu A. Characterization and evaluation of reproductive performance in Bapedi sheep breed. Bloemfontein, South Africa: University of the Free State; 2019

[8] Ramsey K, Harris L, Kotze A. Landrace Breeds: South Africa's Indigenous and Locally Developed Farm Animals. Pretoria, South Africa: Farm Animal Conservation Trust; 2000. pp. 38-39

[9] Snyman MA, Olivier JJ, Cloete JAN. Productive and reproductive performance of Namaqua Afrikaner sheep. Karoo Agriculture. 1993;**5**(2):21-24

[10] Qwabe SO, van Marle-Köster E, Visser C. Genetic diversity and population structure of the endangered Namaqua Afrikaner sheep. Tropical Animal Health and Production. 2012;**45**(2):511-516

[11] Food and Agriculture Organization of the United Nations (FAO). Livestock Keepers – Guardians of Biodiversity. Rome, Italy: Animal Production and Health; 2009

[12] Kunene WN. Characterisation of indigenous Zulu (Nguni) sheep for utilisation, improvement, and conservation. Durban, South Africa: University of Kwa-Zulu Natal; 2010

[13] Saaed FF, Zaid NW. Serum and testicular testosterone levels of ram lamb during puberty. Advances in Animal and Veterinary Sciences. 2018;7(2):92-95

[14] Perkins A, Roselli CE. The ram as a model for behavioural neuroendocrinology. Hormones and Behaviour. 2007;**52**:70-77

[15] de Kretser DM, Loveland KL, Meinhardt A, Simorangkir D, Wreford N. Spermatogenesis. Human Reproduction. 1998;**13**:1-8

[16] Eriksson P, Lundeheim N, Söderquist L. Changes in mean scrotal

circumference in performance tested Swedish beef bulls over time. Acta Veterinaria Scandinavica. 2012;**54**:74-82

[17] Castro ACS, Berndtson WE, Cardoso FM. Plasma and testicular testosterone levels, volume density and number of Leydig cells and spermatogenic efficiency of rabbits. Brazilian Journal of Medical and Biological Research. 2002;**35**(4):493-498

[18] Bearden H, Fuquay J, Willard S. Applied Animal Reproduction. Upper Saddle River, USA: Pearson Prentice Hall; 2004

[19] Söderquist L, Hultén F. Normal values for the scrotal circumference in rams of Gotlandic breed. Reproduction in Domestic Animals. 2006;**41**(1):61-62

[20] Burger A, Hoffman L, Cloete J, Muller M, Cloete S. Carcass composition of Namaqua Afrikaner, Dorper and SA Mutton Merino ram lambs reared under extensive conditions. South African Journal of Animal Science. 2014;**43**(5):27

[21] Perumal P. Scrotal circumference and its relationship with testicular growth, age, and body weight in Tho (Bos indicus) bulls. International Scholarly Research Notices. 2014;**2014**:249537

[22] Foster DL, Hileman SM. Puberty in the sheep. In: Knobil and Neill's Physiology of Reproduction. Cambridge, UK: Academic Press; 2015. pp. 1441-1485

[23] David I, Kohnke P, Lagriffoul G, Praud O, Plouarboué F, Degond P, et al. Mass sperm motility is associated with fertility in sheep. Animal Reproduction Science. 2015;**161**:75-81

[24] Rege JEO, Toe T, Mukasa-Mugerwa E, Tembely S, Anindo D, Baker RL, et al. Reproductive characteristics of Ethiopian highland sheep II. Genetic parameters of semen characteristics and their relationships with testicular measurements in ram lambs. Small Ruminant Research. 2000;**37**:173-187

[25] Ramu S, Jeyendran RS. The hypo-osmotic swelling test for evaluation of sperm membrane integrity. Methods in Molecular Biology. 2013;**927**:21-25

[26] Yarney TA, Sanford LM. Pubertal development of ram lambs: Physical and endocrinological traits in combination as indices of post-pubertal reproductive function. Theriogenology. 1993;**40**:735

[27] Colas G. Factors Affecting the Quality of Ram Semen. London: Sheep Production. Butterworths; 1983. pp. 453-465

[28] Dombo MH. Seasonal effect on semen quality of Gorno Altai and South African indigenous goats. Pretoria, South Africa: University of Pretoria; 2002

Chapter 2

Importance of Minerals in the Diet of Cattle in the Tropical Climate of Mexico

Marco Antonio Esteves Luna, M.C. Samuel Tabe Roldan, Leonel Avendaño Reyes, Fredy Minor Perez, Arnoldo González-Reyna, Francisco Javier Trejo Meza and Julio Cesar Camacho Ronquillo

Abstract

The chapter addresses the importance of minerals in the diet of cattle in a tropical climate to maintain their body condition in relation to productive activity and especially reproductive activity, this importance falls on the physiological processes within the animal organism considering that fantastic world that explain the phenomena that occur within the organism that are important processes since they maintain cellular homeostasis, this allows the generation of cells of the body structure as well as male and female gametes that influence the conservation and maintenance of biochemical phenomena with the emphasis on conserving the organic status that, when unbalanced, generates fertility problems, affecting above all milk and meat productivity, generating economic losses in cattle ranches in southern Mexico. It is convenient to indicate that Mexican soils in a great proportion present deficiencies of minerals such as zinc, selenium, boron, chromium, a few to comment that along with forage production, this is accompanied to produce infertility phenomena in cattle. in grazing that influence the presentation of heats, corresponding to artificial insemination as well as problems in ovarian pathologies that inhibit ovarian cycling and become unprofitable, for this reason this the minerals in the diet is important so that the profitability of milk and meat benefits the producers of the countryside in southern Mexico

Keywords: minerals, nutrition, production, reproduction, bovine cattle, physiology, metabolism

1. Introduction

Ruminants are characterized by their ability to feed on pastures and forages, since they degrade carbohydrates, such as cellulose, hemicellulose and pectin, these are very indigestible for non-ruminant or simple-stomached species. Therefore, the digestive physiology of the ruminant acquires particular characteristics due to the fact that the degradation of the food is carried out mainly by fermentative digestion and not by the action of digestive enzymes, the fermentative processes take place by different types of microorganisms that the ruminant houses in its stomach diversity. However, to reach these stages, the digestive system of bovines undergoes various transformations from birth to the development of the forelimbs, particularly the rumen, being essential to carry out these fermentative processes, which are accompanied by microbial action to facilitate the digestion process. The rumen is the proventriculus with the largest physical size and its physiological importance lies in the fact that the forages are fermented here to make nutrients absorbable and facilitate the extraction of chemical compounds to be distributed in the bloodstream to the liver and then sent to the whole organism. Rumen development is favored by the food ingested, in the first week of life from 0 to 4 weeks they basically depend on the glucose contained in milk [1], the newborn of this age cannot digest pastures or concentrates, in practice, adding 100 grams of concentrate daily to the newborn's diet favors the development of rumen papillae for when they mature, facilitating the absorption of volatile fatty acids (AGVs), thereby facilitating the increase in weight and body size, the AGVs refer to butyric, propionic and acetic acid they enter different metabolic pathways to produce glucose, acetic acid o It quickly crosses the rumen wall without undergoing any change and is used by the organism as an energy supply. Propionic acid is converted into lactic and succinic acid so that it directly enters the Krebs cycle for energy or is used as a glucose precursor. Butyric acid is metabolized in the rumen wall and converted to β-hydroxybutyrate [2]. For [3], forage consumption affects the rapid development of the rumen in size and functioning, after 60 days of birth, the rumen experiences its maximum growth, reaching proportions close to those of the adult with respect to the other digestive organs and the body weight. By this time the calf will be ready to be weaned and the rumen microbiota should be fully developed [2]. Likewise, [4] endorses that the diet is the fundamental or determining factor for the development of the morphology of the rumen wall and the papillary development, for which reason the consumption of balanced foods stimulates the growth and the presence of products that originate AGVs that are necessary factor for papillary maturation. Consequently, with the body development of the calf, the rumen matures, facilitating the growth of bones and body mass, which, supported by the genetic characteristics of each breed, favors meat production. Minerals are

Macrominerals	**Microminerals**
Calcium	Cobalt
Phosphorus	Copper
Potassium	Fluorine
Magnesium	iodine
Sodium	Iron
Sulfur	Manganese
Chlorine	Molybdenum
	Selenium
	Zinc

Table 1.
Macro and micro minerals.

important elements since they are essential nutrients for all domestic animals and influence the productive and reproductive efficiency of almost all species, they constitute about 5% of the total body weight and are involved in most of the metabolic reactions that occur at the intra and extra cellular level. The generalized classification divides them into two large groups: microminerals and microminerals or trace elements as shown in **Table 1**.

2. Developing

Cattle farming in the tropical climate of Mexico presents serious production and reproduction problems due to geographical, technical and social factors, the exploitation system that prevails is extensive livestock, in which the following pastures predominate: 1. Elephant (Penisetum purpureum), Guinea (Panicum maximum), Buffel (Cenchrus ciliare), Para (Brachiara mutica), African Star (Cynodon plectostachyus) and other varieties such as Insurgentes (Brachiaria brizanta), Signal (Brachiaria decumbens), Privilegio (P. maximum) and Bemuda (Cynodon dactilon) despite the great forage production that is achieved during the rainy season, the maximum weight gain is less than 500 grams per day, deteriorating in the dry season due to the lignification of the plant produced by high temperatures, the protein, energy, vitamin and mineral supplementation becomes necessary but due to the high costs it is not used, although some producers use common salt to meet the needs of cattle, mineral supplementation is an alternative, in addition to being low cost, helps to improve the productive and reproductive parameters of animals in the southern region of Mexico. The tropical zones of Mexico represent an alternative for livestock production since pastures are the most abundant and economical source of food to feed livestock. However, there are times throughout the year where the productivity and quality of pastures are affected by climatic conditions, especially temperature [5], affecting the weight gain of grazing animals [6]. The types of soils that predominate are the so-called chronic rainy, lithosols with medium textures with good permeability, efficient superficial and internal drainage, rich in potassium (K) in the first layer and with low amounts of phosphorus (P), calcium (Ca), magnesium (Mg), slightly acidic and free of sodium (Na).

Mineral nutrition in any organic (plants and animals) and inorganic (water and soil) life system is of vital importance for the development of human life [7]; however, the great imbalance that has been generated in the inorganic environment due to the excessive use of fertilizers and chemical substances has caused a general deterioration in both life systems, to such an extent that a mineral disorder has been caused that, as a consequence, causes high deficiencies. and/toxicities [8]. It should be remembered that the interaction that exists between the soil–plant–animal, as a very complex small food chain, if the law of ecological life complies with its rule, only 10% of each link will contribute nutrients to the next trophic level, that is, that only 10% of the soil nutrients will be taken up by the plant and the other 10% of the plant will be taken up by the animal [9]. Although protein and energy inputs represent the main limitations in animal production in the rangelands of southern Mexico, mineral deficiency increases the problem, since these exert a marked influence on the use of these two nutrients [10], so the action becomes reversible, that is, good nutrition is not only considering the contribution of energy and protein, minerals even when they are incorporated in very small quantities are necessary in the daily consumption of the animal to maintain its good physiological state, metabolize and general health of the organism [11].

Minerals fulfill various basic functions within the body, both at the cellular level and in tissues and organs, so a deficiency or excess of one or more of them can cause metabolic dysfunctions that are reflected in low productive and reproductive performance, even in the death of the animal [12]. In the grazing system, unlike the feedlot one, the latter always supplement the bovine diet with complete rations that almost always include a mineral premix, not so in extensive grazing bovines and in tropical environments, where there are generally deficiencies or toxicity of elements. Inorganics that cause negative effects on the growth rate, low meat and milk production, fertility disorders in the animal [10], inadequate use of forage and low digestibility of the same [13], which increases the fattening period from 3 to 5 years.

The most commonly used sources for grazing water supply are small streams, lakes, puddles or springs, wells, and rainwater [14]. All the essential minerals considered as dietary nutrients in the animal are present in the water; however, concentrations for grazing animals are completely inadequate to meet the animal's requirements [14, 15]. The consumption of water for the animal is a function of the food, the content of dry matter in the forage, the physiological state and climate, as well as the quality of the water [14]. For [16], points out that the soil pH influences the root absorption of minerals, acid soils often produce forages with deficiency of Ca and Mb. [17] pointed out that in calcareous soils the availability of phosphorus is related to the content of carbon dioxide and the pH, soils of 6.6. at 7.7 they are more favorable to contain P, being the acidic soils of the tropics marginal in this element [18].

Mg is another mineral that has low availability in crops, due to its low availability in the soil or by interaction with Ca and P, Mg is one of the minerals most sensitive to pH changes in alkaline soils it precipitates and in acids is soluble and available. Fleming [19] comments that acid soils decrease the availability of Fe, Mn, Zn, Cu and Co.

Organic matter is important in relation to the contribution of nutrients from the soil to plants, texture is also relevant in the contribution of nutrients to the plant since it facilitates water retention and makes available Co, Cu, I than in sandy soils. It will be of low availability and in clay soils high in these elements [11].

The nutritional value of tropical pastures is characterized by its low digestibility and nutritional value compared with pastures in temperate climates, the protein increases as the crude protein content increases, this in tropical pastures varies between 50 and 60, likewise, grasses in tropical climates have a high growth rate in rainy seasons considering their medium to good nutritional value at this time, but it decreases rapidly due to the effect of maturation caused by the high temperatures existing in the region [5, 15].

There are several factors that intervene in the availability of an element of the plants as they are; the plant species, the period of growth, the climate and the stocking rate. Within plant species, legumes and grasses are distinguished, the first being rich in Ca, while in pastures it is Mg, although these are deficient in protein, P, Mb, Se and I [7, 16]. In forages in tropical climate, minerals are affected by the growth stage; Na, P, and K decrease as the plant grows and matures, an effect due to leaf drop and decreased absorption of nutrients from the soil to the plant. The climate also affects the presence of minerals because tropical forages have a high photosynthetic rate, light and temperature accelerate the photosynthetic process, making the plant mature soon, increasing the content of the cell wall, on the other hand, McDowell 1977, reported that the solubility of minerals in the soil is diminished in regions with high rainfall, due to loss of edaphic leaching, as it is in the humid and sub-humid tropics.

The stocking rate influences the predominance of certain plant species and changes the leaf/stem ratio, which is why it has a direct relationship with the mineral content of the pasture because the leaves are richer in minerals than the stems, the mineral content is removed in the plants the more there is a greater grazing and increase in the production of dry matter [20–23], reported that continuous grazing produced significantly lower levels of Ca, Mg and Cu compared to rotational grazing, there being no differences in their P content in both systems. The minerals are also beneficial for rumen microorganisms [24], the Co serves as food for bacteria that produce vitamin B12, the inorganic S is only used to stimulate the digestion of cellulose within the minerals involved in the digestion of cellulose are Ca, P, K, Mg, S, Mn, Fe, Cu, Co, Zn, Se and I [13].

The minerals play an important role within the enzymatic system, they form the commonly called Metalloenzymes and enzymes activities by minerals in the first group the mineral is an integral part of the molecule and cannot be removed by dialysis, this type of enzymes generally contain Fe, Cu, Zn, Mb, Mb, Se in some cases have two minerals such as cytochrome oxidase (Fe and Cu) or xanthine oxidase (Fe and Mb), the second group of minerals are not part of the molecule and can be removed by dialysis so that these compounds are unstable among the mineral elements that activate enzymes are; Ca, Na, K, Co, Fe [25]. In the same way, minerals have a direct and indirect effect on the structure and function of hormones.

Within bovine reproduction, minerals play an important role since the scarcity of these increases the periods of infertility, quality and embryonic death, fundamentally marking the deficiencies of the same during the transition period since inadequate diets in elements such as Zn and P it favors the appearance of anovulatory cycles or delays in ovarian cycling that lead to economic losses since the inter-delivery interval is prolonged, losing productive and reproductive life of grazing cattle.

3. Function and general characteristics of minerals: Calcium (Ca) and Phosphorus (P)

Calcium and phosphorus have vital functions in almost all body tissues and must be available to animals in the right amounts and ratios. These elements represent more than 70% of the total minerals in the body and are generally studied together due to the interrelationship between them.

Approximately 99% of the Ca and 80% of the P in the body are present in the bones and teeth.

It has been established that the Ca: P ratio and the level of vitamin D can affect the utilization of both minerals.

Calcium constitutes about 2% of the total weight of the animal. The soil concentration varies in different mammals from 9 to 15 mg/dl. This mineral is in a dynamic state and is constantly deposited and mobilized from the reserve organs. Calcium exists in 3 states in the blood and body fluids.

As ionized soluble Ca (free ion).

As non-ionized organic acid Ca such as citrate.

Protein-bound Ca whose binding increases with ph.

Only ionized soluble Ca is physiologically active.

4. Absorption

Calcium is absorbed by the mechanism of active transport throughout the intestine, although the preferential site of absorption is the duodenum, the amount of Ca absorbed depends on several factors, among which are:

- The form of Ca intake level.
- Absorbed levels of Ca and P.
- The vitamin D status in the animal.
- The presence of lactose.
- Bile salts and fats.
- Phytic acid, iron and oxalate.

Calcium in foods is usually present as salts of phytic, phosphoric, oxalic, carbonic, or tartaric acids. These salts are soluble at low pH and accelerate the absorption of dietary acids.

The ratio of Ca to P has a great impact on the degree of absorption of the blood levels of both elements. An excess of either causes an increase in fecal excretion by formation of insoluble Ca that is not available for absorption.

It has been proven that vitamin D increases the absorption of ca in the intestine only when the salts are soluble. The main hypotheses state that:

- Vitamin D or its derivatives enhance the diffusion of calcium ions through the intestinal wall by counteracting the factors that reduce calcium concentration (Ca + 2) or by increasing the permeability of the intestinal epithelium membrane.
- It is essential for the formation or initiation of the special mechanism of ca transport in the intestinal wall. It has been shown that sugars that are absorbed relatively slowly, such as lactose, ribose, sorbose, xylose, and fructose, can increase ca production by favoring its binding with the protein that serves as a transporter.
- Under normal feeding conditions fats have a minimal effect on intestinal ca absorption. When fats are hydrolyzed in food, giving rise to fatty acids and these are not absorbed, a union will occur for the formation of insoluble ca salt that is lost in the feces.
- Some organic and inorganic compounds such as iron, oxalate and phytic acid can form insoluble salts with ca and thus affect its absorption.

Another factor that regulates ca absorption is parathyroid hormone (PTH) which raises low blood ca levels and maintains normal levels by mobilizing ca in the bone. A high dose of parathyroid hormone can demineralize bone and lead to hypercalcemia.

5. Functions

The main function of ca is bone formation. It has been shown that the content of this mineral in the body is a function of the live weight of the animal.

The intensity of postnatal calcification depends on the physiological maturity of birth.

It is involved in the action of enzymes involved in blood coagulation.

Decreases cell permeability and maintains physiological permeability and its pores by counteracting the effect of sodium and potassium.

It is required for muscle contraction. In its absence, all types of muscles lose their ability to contract. Excess ca stops muscle contraction.

For the maintenance of the correct degree of neuromuscular excitability and tone.

The main routes of ca excretion are the intestine and the kidneys.

The ca concentration in the urine of calves is above 100 mg/dl because during the first weeks of life in ruminant's ca is excreted mainly through the kidneys and not through the intestines.

There is a relationship between the excretion of Ca and Mg. Magnesium increases ca excretion and vice versa, both compete by a common absorption mechanism.

6. Hormone regulation

Several hormones such as parathyroid (PTH), calcitonin, and 1,25-dihydroxyvitamin D2 and D3 are involved in ca homeostasis during pregnancy and lactation.

Parathyroid hormone regulates the ca mechanism and maintains a constant level of this element in the blood in combination with calcitonin. The active process responsible for the homeostatic regulation of the level of ca in the blood can be of two types.

a. Maintain calcemia, that is, the activity of ca carried out within the limits required for normal physiological functions.

b. Maintain the product of ca and p activities (ACa2+ AHPO4-2) in the extracellular fluid at a level that enables the spontaneous growth of hydroxyapatite crystals in the bone.

Thus, the first line of defense against decreased Ca2* activity is when the amount of phosphate in the serum decreases. This is due to increased excretion by the kidneys through the effect of parathormone. This mechanism is very fast and sensitive but its potentialities are limited. The second line of defense against decreased Ca + 2 activity, which is less sensitive but practically inexhaustible, is the mobilization of ca ions from the bones by the direct effect of parathormone. At the same time, the elimination of excess phosphate through the kidneys takes place.

Increases the concentration of ca and decreases the concentration of phosphate ions in the plasma. Thus, it compensates for the effect of the release of phosphate ions during the elimination of ca through the bones. Accordingly, the increase in plasma ca concentration proceeds according to the formula $P \times Ca = K$.

7. Excess calcium

Feeding with high levels of ca reduces feed intake, weight gain and delays sexual maturity. Excess ca can also cause mineral imbalance by chelating other elements such as zn, which reduces ca and causes parakeratosis in pigs. In dairy cows, the incidence of paresis at calving can increase and diets rich in this element hinder the absorption of p, cu and mn.

8. Mineral excretion

8.1 Phosphorus

Phosphorus is excreted in the feces and in almost insignificant amounts in the urine in ruminants. In pigs, they are eliminated in large quantities by the kidneys and the intestine.

Determination of endogenous losses is important in ruminants since these animals excrete endogenous p almost exclusively through the digestive tract and the amount may exceed that present as undigested in the feed.

8.1.1 Availability

Phosphates are found in nature in the form of organic compounds and are used depending on the species and age of the animal.

Phytic acid salts (phytates), especially those of ca and mg, are not digested by some animals, particularly in monogastric species, and assimilation is low. In pigs, a small part of the phytates is hydrolyzed in the stomach by plant phytases, while in ruminants, phytate hydrolysis occurs in the stomach by bacterial phytases. This is explained because the optimum pH for cereal phytases is approximately 5.1 and it shows some activity at a pH lower than 3.

Therefore, there is considerable breakdown of dietary phytate in the crop of chickens and in the stomach of non-ruminants, before gastric secretion reduces the pH of the ingestion to a level too low for nutrient activity. This is not the case in ruminants, because the pH of the fore-stomachs is higher than 5.1.

Phosphorus from inorganic supplements and foods of animal origin can be used up to 100%, while that from plants can range between 30 and 60%.

The digestibility and therefore the availability of p is reduced when diets low in energy are supplied, for which reason it is considered that this is essential to guarantee a good use of p.

8.1.2 Function

In the metabolism of fats, the intermediary metabolism of formation of cephalin and lecithin's (phosphatides) appears.

It is an essential component of nucleoproteins and nucleic acids such as RNA and DNA.

It plays a vital role in carbohydrate metabolism in the formation of hexoses and trioses, as well as energy compounds such as adenylic acid and creatinine phosphate.

Involved in protein metabolism (formation of phosphoproteins), and in muscle metabolism.

Phosphate as a constituent of energy-rich phosphates is of key importance in the energy metabolism of cells (as ATP) a large number of coenzymes are phosphorous compounds.

Phosphorus deficiency is the most widespread in grass-fed domestic animals and is more frequent in cattle than in sheep, since the latter has energy needs per unit weight greater than those of cattle due to its size, smaller consumes more food.

Sheep meet their physiological needs for p with diets whose concentrations are lower than those of cattle. Calcium deficiency rarely occurs in cattle and sheep, except in high production cows that require large amounts of this element.

8.1.3 Phosphorus deficiency leads to

- Poor reproductive performance.
- Inactive ovaries delay sexual maturity.
- Low conception rates.
- Long DAs.
- Embryonic death, calves born weak.
- Excess: endometrium susceptible to infection [26].

8.2 Magnesium

60% of the mg is found with the p-ca as complex salts in the bones and teeth. The rest of the mg is found in tissues and in body fluids. The mg content, unlike ca and p, remains almost constant in the tissues when the animal is an adult. The levels in the soils oscillate between 1.3–3.3 mg/dl and are directly related to their content in the diet.

The assimilation of mg in animals depends on the degree of endogenous loss of this element in feces.

Endogenous mg is secreted in the gastrointestinal tract with saliva and other digestive tracts and can pass through the intestinal wall. The concentration of $Mg + 2$ in the saliva of ruminants is 0.4–0.6 meq/l and varies inversely by the rate of disaggregation.

The endogenous loss in cattle is 3–4 mg/kg LW/day, in pigs, sheep and horses it is 2.5, 2 and 2.3 mg respectively.

In all species the assimilation of mg is low and depends on the type of food. The mg assimilation of adult ruminants is 25–30% in hay, 16–20% in concentrates, 20–25% in mixed diets and 50–55% in other diets.

Magnesium is mainly eliminated in the feces. Most of it is filtered through food and is reabsorbed in the renal tubules. Under normal conditions, endogenous and unabsorbed magnesium is eliminated from the body mainly through the gastrointestinal tract. The amount of magnesium excreted in the urine is relatively small, although this pathway plays a definite role in maintaining magnesium homeostasis. Thus, if the concentration of magnesium in the diet increases, its relative excretion in the urine increases, but up to a certain limit.

Metal ions are found in 2 forms in enzymes.

1. Metalloenzymes in which magnesium is bound to protein.
2. Metalloenzymatic complexes not bound to proteins.

In many catalyzed reactions, magnesium can be substituted for manganese. Another function of magnesium is to influence central nervous system irritability by activating cholinesterase, which breaks down acetylcholine.

8.2.1 Calcium magnesium relationships

The levels of ca and p in the diet have a marked effect on Mg requirements. By raising ca and p, mg needs are increased.

A deficiency of mg in the diet of layers produces a rapid decrease in laying, hypomagnesemia in the blood and depletion of mg in the bones.

Altered mg metabolism produces meadow tetany in herbivores. This disease occurs at the beginning of spring when cattle are taken out to graze on grass that has been fertilized with k or n. The increased production of ammonia in the rumen causes the deduction of mg absorption. The most important biochemical alteration is the existence of subnormal levels of mg and ca in the blood of 1–7 mg/dl of mg and 6.6 mg/dl of ca. Hypomagnesemia can also occur in calves fed dairy diets.

Another disease associated with mg deficiency is puerperal paresis or milk fever that, although the characteristic is acute hypocalcemia, it is also observed that the levels of inorganic p in the serum decrease to half of normal. Magnesium deficiency is usually not appreciated unless severe natural or induced diuresis occurs.

8.3 Sodium (Na), Chlorine (Cl) and Potassium (K)

Sodium, k and cl are the main determinants of acid–base balance and water balance (sodium, alkaline potassium, acid chloride). Acid–base balance and imbalance can affect many bodily functions, including growth rate, feed intake, protein metabolism, bone metabolism, and stress response.

Sodium and cl are extracellular ions. Potassium is intracellular and forms the basis of body cells.

Carnivorous animals can get enough na in their diet. Herbivores, however, should supplement their diet with na, since vegetables generally have low amounts of this element and are rich in k. This causes increased excretion of na. Potassium salts significantly increase the k content in plants. Therefore, there is often an unfavorable k-na ratio of 10:1 and even higher in cattle feed.

8.3.1 Absorption

The site of absorption of Na, K and cl is in the small intestine.

8.3.2 Excretion

The route of elimination of Cl, Ca and K is through the kidney.

The homeostatic mechanism that regulates the metabolism of na and k is sufficiently developed so that animals can survive for long periods with low na intakes and conserve k, except in pathological conditions.

Potassium is practically 100% assimilable. Almost all the K that is eliminated from the organism is of endogenous origin. The percentage of K excreted in the urine is 75–86% in cows, 85–88% in sheep. Neither the amount of K ingested nor the level of feeding affect the ratio between the amount of K excreted by the kidneys and by the intestine.

8.3.3 Biochemical functions

Chloride and na facilitate regulation of osmotic pressure.

Sodium regulates acid–base balance and fluid volume.

Sodium and K increase nervous irritability. In this function they oppose the effects of Ca and Mg.

Chloride is involved in the transport of CO2 by displacing Cl-.

Potassium facilitates the uptake of neutral amino acids.

Sodium is involved in muscle cells during contraction and is necessary for the transport of amino acids, glucose through mucous membranes and cell membranes.

The adrenal gland is important for the regulation of na retention and its insufficiency leads to a reduction in the level of sodium in the blood.

The metabolism of Na in the body is controlled by the endocrine system (mineralocorticoids-aldosterone and oxycorticosterone). Aldosterone controls the process of reabsorption and Na + in the kidney tubules. Na + (and water) retention is usually accompanied by intense K secretion in the urine. Hydrogen ions (H+) that are secreted in the urine compete with K ions (K+). Na reabsorption can be accompanied by a preferential secretion of K+ (in ruminants) and H+.

On the other hand, aldosterone secretion can regulate, by itself, the levels of Na + and K+ in the blood. Aldosterone has been found to suppress mg absorption through the intestinal wall in vitro and to produce hypomagnesemia in vivo. The effect of aldosterone on Mg metabolism may be secondary and is related to mg metabolism of Na and K.

Low serum mg levels were observed in ruminants fed sugarcane straw-based diets treated with na hydroxide.

8.3.4 Deficiencies

Dietary na deficiencies can occur during lactation. This may be caused by the loss of this element in milk in animals that grow rapidly and consume cereals low in Na in warm areas where there are extensive na losses through sweat. It also occurs in animals subjected to intense work and in animals that consume pastures deficient in this mineral.

Farm animal diets are often deficient in na and for this reason the level of the element in the ration must be constantly monitored.

Deficiencies and excesses of k are rare in animal feed.

Sodium deficiency affects bovine reproductive physiology.

8.4 Sulfur S

Sulfur represents 0.15% of the body weight. This element is present in all the cells of the body, mainly in organic tissues rich in protein. Sulfur and n metabolism are associated, since two important amino acids cystine and methionine contain sulfur.

There is evidence that s may be limiting in certain diets. With the increase in the use of urea in the ration, an increase in inorganic sulfate supplementation is suggested. The s requirement of the animal is covered with the content of s-containing amino acids and, partially, with heterocyclic compounds such as biotin and thiamine.

8.4.1 Absorption

The absorption of s occurs mainly in the small intestine. Among the different sources, inorganic s is absorbed less than organic s in some species, although not in all.

8.4.2 Excretion

Sulfur is eliminated in feces and urine, depending on how you administer the amount received.

Possible use, by microorganisms in the digestive tract, of elemental s or sulfate that is added to diets to cover the deficiency.

Sulfur is essential for the synthesis of certain compounds (mucopolysaccharides) in the body.

Sulfur is essential for microorganisms, for the digestion of cellulose, utilization of n-p-n (non-protein nitrogen) and for the synthesis of B complex vitamins.

8.4.3 Deficiencies

In ruminants, s deficiency can occur if protein is replaced by non-protein nitrogen (NPN).

8.4.4 Excess sulfur

Excess mineral s as sulfate has an adverse effect on chickens and piglets (growth inhibition, gastroenteritis). Caution should be exercised when using Clauber's salt as a source of s and na in calf and cow diets.

8.5 Manganese (Mn)

It appears distributed throughout the body in minute amounts, although it tends to concentrate in the mitochondria. It is also found in the bone, liver, pancreas, kidney, brain, and heart.

Bone is the main source of mn and can serve as a deposit of this element.

It is also found in the liver, muscles and kidneys. The concentration of mn in the hair or feathers can be correlated with the level in the diet and it has been recommended that the content in the body hairs of pigs be taken as a criterion of an adequate supply of the diet.

8.5.1 Absorption

Manganese is absorbed mainly in the small intestine, specifically, in the duodenum in polygastric and monogastric animals, and its absorption decreases when there are excessive amounts of calcium, phosphorus or iron. Absorbed MN is rapidly cleared from the blood to the liver, bones, and hairs.

8.5.2 Excretion

Manganese is eliminated through the bile and into the intestinal tract and is excreted in the feces. It is absorbed as bile-bound mn. Each atom crushes several times before its final excretion. The excretion of mn in the urine is negligible.

As part or activator of numerous enzymes such as the enzyme arginase.

In the function of the skeleton, the muscles and the development of the genitals.

It is essential for the development of the organic matrix of bone, which is composed of mucopolysaccharides.

It is required in the synthesis of fatty acids as a mn chelate.

It is involved in the metabolism of amino acids with pyridoxal phosphate.

8.5.3 Deficiencies

Although mn deficiencies have been produced experimentally in many animals, it has only been presented as a practical problem in the feeding of birds whose dietary needs are higher than mammals [26].

Deficiencies manifest as growth retardation, skeletal abnormalities, impaired reproductive functions, and ataxia of newborns. In bovines, a deficient diet in mn delays sexual maturity and decreases fecundity. Milk production decreases.

8.6 Zinc (Zn)

The normal animal organism contains a Zn concentration of approximately 30 parts per million. The highest concentrations of zn in the body are found in epidemic tissue such as skin and hair, although they have no preference for any particular tissue.

8.6.1 Absorption

In most species, zn is mainly absorbed in the upper segment of the small intestine (less than 10% of the amount ingested). Absorption varies according to how the item is ingested. Zinc carbonate, sulfate, and oxide, as well as metallic Zn, are absorbed in the same way.

In the rumen of ruminants that receive grass diets there are only 5–10% of zn in the soluble form (grass has 50%). It appears that Zn binds to the microflora of the foreskin. In the abomasum and duodenum the solubility increases and can be more than 80%. Excess phytic acid reduces the absorption of Zn. White tissues have a large exchangeable zn deposit that maintains a balance with plasma Zn. Organs such as the liver and renal cortex are rich in mitochondria and have a maximum amount of Zn.

Zinc is eliminated mainly in the feces and less than 5% of the ingested is eliminated in the urine. There is an effective homeostatic mechanism for Zn in the intestine. Homeostasis is maintained by variations in the amounts of Zn absorbed and its endogenous excretion in feces.

It is an essential component of carbonic anhydrase, an enzyme that plays an important role in the basic balance in the body and in the release of CO2 in the gastric mucosa.

8.6.2 Deficiencies

In calves they are manifested by subnormal growth, alopecia, parakeratosis in some regions of the body. The lambs show appetite disturbances, eat wool and reduce growth.

Biochemical alterations in blood and tissues are not constant, but as deficiency progresses there is a slight decrease in Zn in liver, kidney, heart and muscle tissues and a more intense decrease in the pancreas.

Zinc deficiencies also lower the blood level along with alkaline phosphatase.

Interrelation of Zn with other minerals. High Zn intake decreases cu and fe retention. The interaction between Cu and Zn was forced by seeing that Zn provides good protection against Cu poisoning.

There is an interrelationship between the ingestion of Zn and Ca, since the main parakeratosis has been observed and in pigs that receive diets high in Ca.

8.7 Iron (Fe)

In adult animals it is found in amounts ranging from 60 to 90 ppm in fat-free tissues. Approximately 57% of the total fe is in the hemoglobin of the blood and 7% in myoglobin. Iron is stored for the most part in the form of ferritin and hemosiderin.

The bone marrow is one of the last reserves that can be depleted of iron and also one of the last to recover.

The duodenum is the major site of fe absorption in the digestive tract, although some absorption occurs in the stomach. This is absorbed in the form of ferrous ion and is reduced to ferric in the stomach. The absorption mechanism is believed to be as follows, ferrous iron enters the mucosal cell and is oxidized to ferric form. This combines with the protein apoferritin to form ferritin, which requires energy from phosphate bonds. At the other end of the cell, Fe is reduced to the ferrous state, separating from ferritin, it passes into the blood and, after autoxidation and in the presence of CO2, it binds to siderophyllin to be transported as ferric iron.

One of the main functions of Fe is to be a component of heme that combines with qlobin to form hemoglobin. It acts as a component of cytochromoxidase and xanthine oxidase. Muscles contain an oxygen-carrying compound, iron-containing myoglobin.

The first sign of deficiency is microcytic hypochronic anemia caused by insufficient fe for normal hemoglobin formation.

There is no convincing evidence that animals grazing under natural conditions suffer from Fe deficiencies. Young animals of any domestic species can be Fe deficient if they eat a dairy diet and lack other external sources of Fe.

8.7.1 Excesses

Many times the excess of Fe salts of food origin can cause nutritional disorders, form insoluble phosphate that reduces the absorption of P, which produces rickets, insoluble Fe phosphate absorbs vitamins or inorganic trace elements that prevent their absorption.

8.8 Copper (Cu)

Copper is an essential element for food. In most adult organisms of the species they are found in a concentration of 1.5 to 2 ppm. The highest concentration of cu is found in the liver as dietary levels raise the liver content.

8.8.1 Absorption

It takes place in the upper portion of the small intestine and increases with CIH and decreases with ca. About 28% of ingested cu is absorbed.

Absorbed Cu bound to plasma albumin is rapidly transported to the liver and other organs where it is stored as a Cu-containing protein. The liver is the major storage site for Cu and the status of Cu balance in the animal can best be assessed by determining the Cu content in the liver. A number of factors such as age, growth hormone, and pregnancy influence the cu balance in the animal.

8.8.2 Excretion

Most of the Cu is excreted in the bile. Urinary Cu excretion is associated with the concentration of Cu that is not bound to plasma protein.

Copper is involved in the formation of young erythrocytes, but not in the concentration of hemoglobin. It has a functional role in fe absorption.

Copper participates in the process of osteogenesis; in protective functions, in the pigmentation and keratinization of the hair. It is a component of cytochrome oxidase, ceruloplasmin, galactose oxidase, and uricase.

Deficiency diseases. Anemia is a general symptom for all species. In cattle, Cu deficiency results in high levels of Fe and low levels of Cu in the liver.

Reproductive problems have been detected in cows that consume diets lacking in Cu.

Although animals are not very sensitive to excess Cu, the problem of toxicity has been pointed out in recent years, especially in ruminants.

8.8.3 Biochemical functions

Excessive doses of Cu (including Cu in milk replacers) and the indiscriminate use of Cu.

Use of Cu sulfate to deworm animals.

Use of Bordeaux mixtures in the silage of certain crops.

Presence of pig and poultry excreta (from animals fed high doses of Cu) in ruminant diets).

Early embryonic death, retained placenta, placental necrosis, better conception rates, silent estrus and low fertility.

In males, low libido, poor seminal quality, sterility.

8.9 Selenium (Se)

It is considered an essential element in nutrition and is associated with Se in organic and inorganic compounds.

In ruminants, a large percentage of ingested Se can be incorporated by rumen microorganisms along with cystine and methionine. They are absorbed and deposited in the tissues as Se amino acids.

8.9.1 Absorption

Seleno-methionine is absorbed from the gastrointestinal tract by an active transport mechanism apparently very similar to that involved in the conduction of methionine through the intestinal mucosa.

8.9.2 **Excretion**

It is excreted by several routes, being the greatest losses through urine, feces and expired air.

8.9.3 Metabolic function

It has been proven that small amounts of se stimulate vital processes and counteract some unfavorable effects of vitamin e deficiency.

8.9.4 Deficiency

Its deficiency is related to infertility, miscarriages, retained placenta, and delayed involution.

In calves, se deficiency, together with vitamin E, P and other factors, can be the cause of white muscle disease (muscular dystrophy).

8.9.5 Excess

Prolonged digestion of relatively small amounts produces toxic symptoms, especially in young animals where growth is inhibited.

8.10 Cobalt (Co)

8.10.1 Absorption

It is absorbed in the proximal portion of the digestive tract of ruminants.

In monogastrics it is synthesized and absorbed in the distal region.

8.10.2 Excretion

Its main route of excretion is urine, although smaller amounts are eliminated in the feces.

8.10.3 Biochemical function

The first or perhaps the only function of Co is the synthesis of vitamin B12 and it represents 4% of the molecule.

8.10.4 Deficiency

Deficiency has only been observed in ruminants and can be caused by low co content in soils or by low availability of this element for plants.

Ruminants synthesize vitamin B12 using Co. When this element is scarce, the synthesis of vitamin B12 decreases, causing anemia.

8.11 Iodine (I)

8.11.1 Absorption

It is absorbed throughout the digestive tract and through the skin, and is present in small amounts in the feces, indicating that almost all orally administered is absorbed.

8.11.2 Excretion

It is almost entirely excreted in the urine. Urinary is constant enough to serve as an indicator of the animal's thyroid function.

8.11.3 Biochemical function

It is used in conjunction with tyrosine in regulating the metabolic rate of the organism.

8.11.4 Deficiency

Iodine deficiency in domestic animals results in dead offspring. Those who survive have enlarged thyroids.

9. Conclusions

Finally, we can say that macrominerals, microminerals and trace elements are of vital importance in the body development of grazing cattle because they enter metabolic cycles to strengthen organic anabolism and catabolism. In mineral nutrition there are 15 elements considered essential, 7 macrominerals: Calcium (Ca), Phosphorus (P), Potassium (K), Sodium (Na), Chlorine (Cl), Magnesium (Mg) and Sulfur (S), and 8 microminerals: Cobalt (Co), Copper (Cu), Iodine (I), Iron (Fe), Manganese (Mn), Molybdenum (Mo), Selenium (Se) and Zinc (Zn) participate in the generation of milk and meat but that Mexican soils are deficient in them, so it is important to carry out soil analysis to identify deficiencies that may present and be careful when supplementing animals and avoid losses due to growth delays, especially in presentation puberty and reactivation of the ovarian cycle after the birth. One of the greatest benefits of mineral supplementation lies in the positive effect on the rumen microbial population, having a great impact on production. Rumen microbes use Ca, P, Mg, K, Fe, Mn, Co, Cu, Mb and S as non-protein N sources, facilitating protein synthesis, fiber digestion, carbohydrate fermentation and energy generation. The minerals cause antagonistic effects to other minerals and are toxic to animals. For example, Cu poisoning is caused by low mo intake, causing hemoglobinuria, gastroenteritis, and sudden death. The loss of Ca-P balance (2:1) is characterized by the presentation of irregular heat or long periods of anestrus [27]. Na intoxication can cause cerebral edema and polyencephalomalacia. The imbalance in Mg, Ca and P can generate urolithiasis. According to the nutritional requirements of cattle, cattle need at least 17 minerals as indicated by the National Academies of Sciences, Engineering 2016.

Author details

Marco Antonio Esteves Luna[1*], M.C. Samuel Tabe Roldan[1], Leonel Avendaño Reyes[2], Fredy Minor Perez[3], Arnoldo González-Reyna[4], Francisco Javier Trejo Meza[5] and Julio Cesar Camacho Ronquillo[6]

1 National Institute of Technology of México, Altiplano Campus of Tlaxcala, México

2 Institute of Agricultural Sciences of the University of Baja California, México

3 Official General Baccalaureate "Land and Freedom", Tlaxcala, México

4 Autonomous University of Tamaulipas, México

5 Jorge Arnáez Genetic Improvement Center of the Tamaulipas Regional Livestock Union, Mexico

6 School of Veterinary Medicine and Zootechnics of the Meritorious Autonomous University of Puebla BUAP, Mexico

*Address all correspondence to: mae1071260@hotmail.com

References

[1] Blanco Ochoa MA. Gastrointestinal development in calves II. 2021. Available from: https://bmeditores.mx/ganaderia/desarrollo-gastrointestinal-en-becerras-ii/

[2] Bacha F. Neonatal calf nutrition. XV FEDNA Specialization Course: Advances in Animal Nutrition and Feeding. Spain; 1999

[3] Trivino PA. Evaluation of Two Feeding Systems in Rumen Development and Productive Response of Artificially Reared Calves. Chile: Thesis. University of Valdivia; 2010

[4] Correa A, F. Study of the Development of the Stomachs of Ruminants. Cuba: College of Veterinary Medicine University of Granma; 2006

[5] Minson DJ. Forage in Ruminants Nutrition. New York: Academic Press Inc.; 1990. pp. 85-149

[6] SAGAR. Statistical Yearbook Oaxaca delegation Secretary of Agriculture. Oaxaca Mexico: Livestock and Rural Development; 1997

[7] Pfander WH. Animal Nutrition in the tropics-problems and solution. Journal of Animal Science. 1971;**33**:843

[8] Stonaker HH. Beef production Systems in the tropics I. Extensive production system on infertile soil. Journal of Animal Science. 1975;**41**:1218

[9] Odum JL. Ecology of the Environment Ecological chains. Mexico: Trillas; 1990. pp. 12-67

[10] McDowell, R. E. Hernández-Urdaneta. 1975. Intensive Systems for Beef production in the Tropics.

[11] Georgievskii VI, Annekov BN, Samokhin VT. Mineral Nutrition of Animals. London: Butterworths; 1982. p. 463

[12] Ammerman CB, Goodrich RD. Advances in mineral nutrition in ruminants. Journal of Animal Science. 1983;**57**:519

[13] Martinez A, Church. Effects of various mineral elements on in vitro rumen cellulose digestion. Journal of Animal Science. 1970;**32**:982

[14] McDowell LR. Nutrition of Grazing Ruminants in Warm Climate. Gainwaville, Florida: Academic Press, Inc.; 1985. pp. 189-311

[15] De Alba J. Livestock Feeding in Latin America. Mexico: The Mexican Medical Press; 1983. p. 475

[16] Underwood EJ. Minerals in Cattle Feed. 2nd ed. Barcelona, Spain: Acribia Publishing House; 1969. p. 325

[17] Murray RB, Mayland HF, Van Soest PJ. Growth and nutritional value to cattle of grasses on cheat Grass range in southern Idaho. Intermountain and range. Experiment station forest service, U.S. Department of Agriculture Ogde, UTHA. Bulletin Int-199. 1978. 57p.

[18] Olson RA, Engelstad OP. Soils of Humid Tropics. Washington, D.C.: National Academy of Science; 1972

[19] Fleming GA. Mineral Composition of Herbage: Chemistry and Biochemistry of Herbage. New York: Academic Press; 1973. p. 529

[20] Carambula M. Production and Management of Planted Pastures.

Montevideo, Uruguay: Editorial Southern Hemisphere; 1977. p. 463

[21] Shute KH. Biology of Microelements and Their Role in Food. Madrid, Spain: Tecnos; 1966. pp. 231-290

[22] Pieper RD. Effects of herbivores on Nutrient Eveling and distribution. In: The Impact of Herbivores in Arid and Semiarid Rangelands. Las Cruces New Mexico: New Mexico State University; 1977. p. 249

[23] Walton PD, Martinez R, A.W. A comparison of continuous and rotational grazing. Journal of Range Management. 1980;**34**:19

[24] Amnenkov. Mineral metabolism in the digestive tract. In: Georgievskii I, Annenkov BN, Samokhin VT, editors. Mineral Nutrition of Animals. Thank Britain; 1981. p. 225

[25] Mertz W. Trace Elements in Human and Animal Nutrition. Fifth ed. Orlando, Florida: Academic Press Inc; 1986

[26] Cordova Izquierdo A, Iglesias Ruiz AE, Ruiz Lang G, et al. Nutrition and food in the reproduction of cattle. European Journal of Agriculture and Food Sciences. 2021;**3**(3):21-33. DOI: 10.24018/ejfood.2021.3.3.184

[27] Valdez-Arjona L, Ramírez-Mella M, Rayas-Amor A, Díaz-Ramírez M, Jiménez-Guzmán J, García-Garibay M. Productive and Reproductive Problems due to Mineral Deficiencies in Cattle from Some Tropical Regions of Mexico. Mexico: AGRO Productivity; 2019

Section 2

Plant Management and Production

Chapter 3

Biofortification of Wheat Landraces: Recent QTL Mapping Advances

Anand Kumar and Prashant Kaushik

Abstract

Micronutrients play an indispensable role in human and animal growth. In the world, many people are suffering from malnutrition and micronutrient deficiency mainly due to lack of zinc (Zn) and iron (Fe). Several crops are grown, such as wheat, rice, maize, and legumes, to address the challenges of micronutrient deficiency. Wheat landraces are evidently proven to be a rich source of genetic variability as against modern cultivated varieties due to thousands of years of their cultivation under low input farming systems. Landraces serve as a potential reservoir of desirable allelic forms of valuable traits but are low in traits like Zn and Fe. Wheat is a major cereal consumed worldwide and could be a good source to provide these micronutrients. Biofortification in wheat can be an effective way to solve the problem of malnutrition. Biofortification in wheat varieties may be enhanced by the application of molecular breeding approaches, such as genome editing, transgenic technology, and marker-assisted selection. These biofortified wheat varieties show better adaptation to environments. In this chapter, we included the recent advances in quantitative trait loci (QTLs) in biofortified wheat and the techniques used to develop biofortified wheat varieties.

Keywords: biofortification, wheat landraces, quantitative trait loci, molecular breeding, micronutrient

1. Introduction

In the world, many people are suffering from malnutrition. Malnutrition occurs due to the nonavailability of proper diet including primary deficiency of zinc (Zn) and iron (Fe) to the living being. Nutritional quality and food quality must be improved by biofortification using the breeding program and it is a sustainable solution to the global malnutrition problem [1]. The Biofortification concept was started at the time of green revolution (1966–1985) [2]. In 1990, the work on micronutrients was started by the American economist by the name of Howarth E-Howdy. But, the term, Biofortification, was coined by Steve Beebe in 2001 [3]. Biofortification provides cost-effective sustainable micronutrients for human beings. This approach will help us to maintain the nutritional status of a living being. In addition, the biofortification process provides micronutrients in wheat varieties, and these wheat varieties are preferred for agronomic practices [4]. These wheat varieties have sufficient micronutrients, especially Zn and Fe. For a better life, Zn is the major micronutrient for human beings as well as plants and its role is observed in both plants and humans [5]. An approximate 17% population has not used up the benefit of Zn [6]. To provide the availability of Zn and Fe, wheat landraces are developed and cultivated globally [7].

Primitive cultivars and landraces are evidently proven to be a rich source of genetic variability as against modern cultivated varieties due to thousands of years of their cultivation under low input farming systems [8]. Landraces serve as a potential reservoir of desirable allelic forms of valuable traits and therefore, could help in biodiversity enrichment and subsequently [9]. A large collection of wheat landraces with enormous variability for different traits still remain within gene banks, without being explored for their utility [10]. With the advancement of modern technologies, applications of molecular breeding approaches, such as genome editing, transgenic technology, and marker-assisted selection, are developing biofortified wheat varieties [9].

Wheat is one of the most popular cereal crops grown and is placed in the second position after rice worldwide and consumed by the people for food and feeding purpose globally [11]. On the earth, the population is increasing alarmingly, however for the purpose of food consumption, wheat production should be increased for food security [12]. Wheat is grown for a better yield under different climatic conditions and is adopted in every diverse area worldwide. It is having to constitute about 10–12% protein and 20% dietary energy. Improving the role of biofortified wheat varieties is the main objective [13]. At the global level, there are many organizations that are playing a key role in biofortification breeding, especially for developing higher zinc and Fe micronutrient varieties.

2. Importance of zinc and iron in humans

In the world, many people are suffering from malnutrition. Malnutrition occurs due to the nonavailability of a proper diet, including primary deficiency of zinc (Zn) and iron (Fe). In the present study, it is observed that the deficiency of zinc and iron is found in dietary food. For a better life, both Zn and Fe are the major micronutrients for humans [14]. The nonavailability of Zn and Fe may be increased using conventional breeding along with molecular breeding [15]. For humans, the availability of Zn was established in 1961 [14]. Human body has a zinc content of about 1.4 to 2.3 g. Because of its nature, its rank is good in periodic numbers in the Periodic Table, and zinc has chemical properties that make it important in biological systems. Zinc plays an indispensable role in cellular function, cell division, cell growth, cellular transport, immune deficiency, transcription, translation, synthesis, platelets, blood cells, immune system, muscles, and liver bones. The quantity of Zn is present in the cell membrane (10%), nucleus (30–40%), and in the cytoplasm (50%) [16]. Zinc also has a good effect on the immune system of a human being.

The deficiency of zinc causes dangerous effects and damage to immunity cellular mediators, such as natural killer cell activity and the production of cytokines [17]. Zinc is an inflammatory and anti-oxidative stress agent. Zn pills may be used for the treatment of diarrhea, and bouts of the condition [18]. Zinc also has an effect on the brain and increases the mind's ability for proper learning through neurons. The common cold may be treated with zinc by reducing the duration and severity of cold in healthy people. Zinc also plays an important role in wound healing, skin creams for treating diaper rash or other skin irritations and may increase the development of sperm in humans [19]. In addition, in 2020, humans faced more problems with Severe Acute Respiratory Syndrome Coronavirus 2 (SARS-CoV-2) (Coronavirus Disease 2019 (COVID-19)) [20].

On the other hand, iron is an essential nutrient; it helps in regulating the flow of oxygen from the lungs to the body tissue. Iron is very necessary for hematopoiesis, regulation of metabolic energy, normal brain development, and muscle development, conversion of blood sugar to energy, and proper growth and development [21]. The deficiency of iron in diet causes severe harmful effects on growth development and muscle metabolism of the body [22]. Therefore, iron is beneficial for a human being, and it differs from gender to gender. Females require more iron than adult males, because females lose blood during the monthly menstrual cycle. However, preadolescent males require more iron and prevalence of higher anemia than preadolescent females. A high quantity of iron is recommended for females aged 14–50 years than the same quantity required for males. This report is given by the Food and Nutrition Board at the National Academies of Sciences, Engineering, and Medicine (USA) [23]. Physiological requirements of iron at the time of pregnancy and lactation show that a high iron requirement is needed during pregnancy and less iron during lactation than in nonpregnant and nonlactating females [23]. Iron is a micronutrient for the proper function of living organisms.

Both iron and zinc play good roles in living beings and respond to regulatory roles of the body as cofactors for producing energy and lipid metabolism [24]. The absence of these elements produces disorder in body involving degenerative nerve cells and disorder in cells.

3. Molecular breeding for quality improvement

Quality improvement is one of the most important aspects of wheat. But the question is how to improve the quality of wheat. A prerequisite that specifies quality of the target gene is needed for the improvement of micronutrients (Fe and Zn) [25]. Getting the knowledge about micronutrient genes Fe and Zn in the crop is a very difficult task and searching where these micronutrient genes in the crop are present in chromosomes [26]. These micronutrient genes depend upon the various physiological, biological, and molecular processes. The main aim of plant breeders is to get the additive effect, transgressive segregation, and heterosis for iron and zinc concentrations in wheat grain [27]. The genetics of iron and zinc's enormous contribution to wheat have been studied. Several studies on wheat landraces have been conducted, including conventional breeding, transgenic, genome-wide association studies (GWAS), clustered regularly interspaced short palindromic repeats/CRISPR-associated protein 9 (CRISPR/Cas9), mutagenesis, genomic selection, etc., for biofortification (**Figure 1**) [21, 28].

3.1 Conventional breeding approach

Conventional breeding approach is a good tool for the improvement of biofortification through conservative manipulation of plant genome in a limited area without harm to the environment. Despite this, conventional breeding approach is less

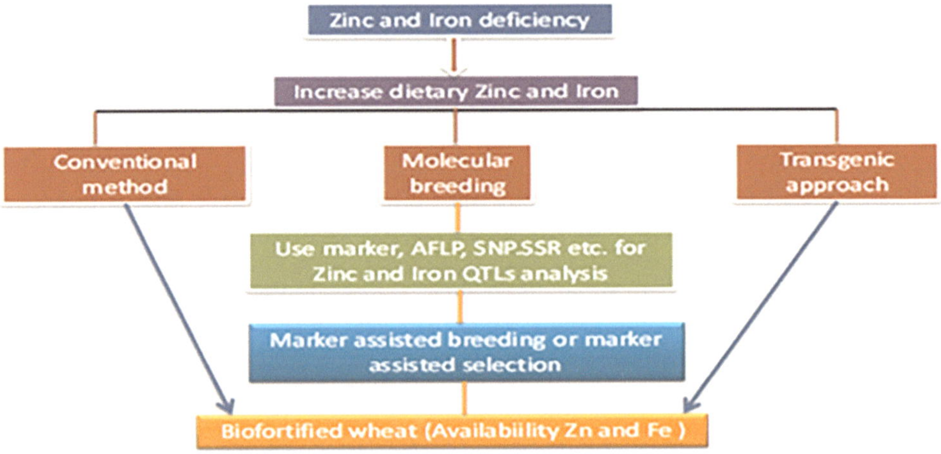

Figure 1.
Schematic representation of the development of biofortified wheat.

preferable [29]. Development of biofortified varieties using traditional methods, such as mass selection, pure line selection, pedigree method, bulk method, backcrossing, single seed descent, synthetics, and composite, is very popular [30], but these methods have some problems, such as taking more time and being labor consumable. In wheat landraces, screenings of germplasm for both yield and quality improvement are a tedious job for a plant breeder, but these aforesaid programs (Genome-wide association studies, CRISPR/Cas9, mutagenesis, genomic selection, etc.) are very useful to overcome problems of a plant breeder [28].

3.2 Quantitative trait loci (QTL) mapping

Another approach to QTL mapping is molecular breeding, which is one of the most important tools by which quality control research work in wheat is done successfully. In this approach, marker-assisted breeding or marker-assisted selection is a very potent tool for bringing about quality improvement in wheat [31]. The QTL mapping in wheat for Zn and Fe has been applied to the population; these populations may be mortal (segregating) and immortal (nonsegregating). The mortal population comprised F1 and backcross and immortal population such as recombinant inbred lines (RIL) (getting after 6–8 generations by single seed descent), and many QTLs have been identified in wheat now [32, 33].

The first QTLs for iron and zinc were identified in 1997 on chromosome 6BS by working on recombinant inbred lines, which were developed by the cross between wheat and wild emmer (*Triticum turgidum*) [34]. This quantitative trait locus (QTL) is the gene *Gpc-B1*, which conferred that the increments of iron and zinc are 18 and 12%, respectively, and that this gene has the NAC (NAM (no apical meristem), ATAF (*Arabidopsis* transcription activation factor), and CUC (cup-shaped cotyledon)) transcription factor NAM-B1 [35, 36]. In addition, several QTLs have been identified for iron and zinc which have been given in the table and some other reviewers presented them elsewhere in their papers [37]. In each study, different QTLs have been identified due to the use of different populations, environments, and marker data sets, which is affected by the location [38]. The use of reference sequence makes it easier to detect the physical position for identification of QTLs with the help of markers on to the genome sequence, and to help in the detection of QTLs for iron and zinc [39]. In these studies, it may be useful to detect the small-effect QTLs for iron and zinc that were not previously possible [40]. However, detecting the QTLs with the help of marker-assisted selection in a large population due to practical limits is a tedious job. However, a small effect of QTLs is being identified using marker-assisted selection from a haplotype and has multiple beneficial QTLs [41]. This haplotype is done as a work of a single unit using a marker-assisted breeding program and reducing the number of markers in later stages [42].

On the other hand, an increment in both traits of iron and zinc may facilitate a significant and positive relationship with several environments; it may have happened by colocalization of the aforesaid QTLs that are underpinned by the pleiotropic effect of a single gene. For instance, by altering the rate of senescence, both iron and zinc content are affected by transcription factor NAM-B1 which takes a step for remobilization of multiple micronutrients [36, 42].

3.3 Identification of a candidate gene for biofortification

In wheat, an annotated chromosome genome sequence has a total assembly size of 14.5 GB which represents the total 94% whole genome and was published in 2018 [43]. The high level of assembly of the genome helps to build up in the genome mapping of loci which are involved in iron and zinc biofortification. In total, 107,891 high-level confidence gene models have been annotated that helped in genome-wide analysis for biofortification of related gene family and identification of loci in QTL. For example, by orthology of rice, the members of the wheat natural resistance-associated macrophage protein (NRAMP) family have been identified by Ali and Borill, 2020 [42] (For further details, see review paper). The NRAMP family illustrates the idea that analyzing a gene family with up-to-date references is necessary. In the previous assembly, several NRAMP gene models were missed out but now they are complete [44]. It has been studied that the RefSeq v 1.1 gene annotation is not perfect while it has significant improvement, for example, 24 NARMP genes were studied, in which two genes were incomplete [42]. This problem is overcome by an ongoing work for improving the annotation of Chinese genome assembly.

3.4 Genome-wide association studies

Even though marker-assisted breeding is a good tool to develop biofortification varieties, the small effect of QTL is not identified by marker-assisted breeding and it also requires a biparental population [45]. Therefore, this restriction is broken by genome-wide association studies. Genome-wide association studies are a powerful tool, because they do not require biparental mating, which is required for QTL mapping [46]. Recently, genome-wide association mapping is a prominent tool that has been utilized by plant breeders, which involves natural variation in elite lines and diverse lines by the production of a dense linkage map, whereas marker-assisted breeding requires a biparental population [47]. In genome-wide association studies, stable marker-trait association and multitraits as well as a large number of genes are responsible for iron concentration in wheat [48, 49]. GWAS have more wide adaptability and are less time consumable than QTL mapping because of which, segregating population is not required. In GWAS, natural variations were exhibited that led to higher resolution mapping than QTL mapping [42].

Moreover, Zhou et al. (2020) used GWAS for increasing the zinc content in wheat with a diversity panel of 207 bred wheat varieties in three locations. In this panel, a total of 29 zinc QTLs that are associated are identified. On chromosomes 1B, 3B, 3D, 4A, 5A, 5B, and 7A, seven nonredundant loci are located in at least two environments. Six coincident known QTLs were identified out of these studies on QTL, in that, 3D chromosome that was previously not identified showed the highest QTL effect in this study [50]. Liu et al. 2019 determined that nine loci are responsible for iron concentration in grain and some other genes were also identified [51]. A total of 35,648 high-quality genotyping and sequencing were used in 123 synthetic hexaploid wheat with the help of single nucleotide polymorphism (SNP) across all 21 chromosomes [48].

3.5 Genomic selection

Genomic selection is another method that is used by the plant breeder for bringing about the improvement of a complex trait like yield in wheat. Applied genomic selection in wheat would accelerate genetic gains in the development of a nutritionally enhanced variety [52]. It also states about the minor gene QTLs. And genomic selection can increase genetic gain by early selection. Velu et al. (2016) evaluated a set of 330 entries from the harvest plus association mapping (HPAM) panel which are derived from diverse wheat genetic resources in which the panel was genotyped with 90 K illumina SNP chip to examine the potential of genomic selection for Zn in two different sets of the environment in India and Mexico [53]. In this, 39 marker-trait associations were discovered using GWAS, and chromosome numbers 2 and 7 showed a larger effect QTL region [54].

Quantitative trait loci and GWAS are more utilized to dissect iron and zinc content in wheat, but these techniques take more time and are labor intensive. In GWAS, only natural variations can be exploited, and we cannot develop biparental mapping [55]. Hence, these techniques are overcome by the transgenic approach. The transgenic approach is a good rapid method to increase micronutrients in grains. In addition, the transgenic approach not only increases the iron and zinc content in grains but also elevates the micronutrient content in the endosperm of wheat [42]. A few studies have been conducted on wheat for increasing transformation efficiency and novel techniques to transform elite lines of wheat [56, 57], which will help in rapid characterization. Transgenic technology will help us to elucidate the function of candidate genes, in that either these genes are identified by mapping approaches or by the ortholog genes. The first transgenic approach was aimed to increase iron not only in grain but also in the storage of iron [56, 57].

3.6 CRISPR/Cas9

Genome editing technology, such as zinc finger nuclease (ZFNs), transcription activator-like effector nucleases (TALEN), and clustered regularly interspaced short palindromic repeats (CRISPR) and Cas9, serve as good and potential tools for biofortification in several crops without altering the whole genome. It alters and knocks out interested genes by the induction of small deletions that lead to the frameshift mutation, whereas it targets only the specific gene of interest [58–60]. Moreover, it is preferred by nongovernmental organizations and governmental organizations, unlike transgenic crops. Therefore, it is highly recommended by eminent plant breeders and scientists. These techniques employed in crops have been widely used for the improvement of some specific characteristics, such as biotic and abiotic stresses and biofortification, in crops [61]. The first-generation technology ZFN and TALEN are less preferred due to low target specificity and labor-intensive nature and develop many off-target cleavages in the genome. While TALEN required a thymine base at starting time that possesses a large size and is repetitive in nature [62, 63].

On the other hand, the use of CRISPR/Cas9 may bring in rapid advancement for Zn and Fe biofortification in elite wheat cultivars and saves the time that is taken by the backcross to remove the unwanted gene that came from conventional breeding. Simultaneously, the editing of homologous genes has been studied; however, a low percentage of transgenic lines have all three homologously edited genes [60]. In addition, little work has progressed using CRISPR for biofortification.

3.7 Mutagenesis

Mutagenesis is an alternative route to manipulation of gene functions, but to find the gene of interest in genome using target-induced local lesions in genome (TILLING) is a tedious and laborious work for plant breeders. Sequence mutant population is developed in hexaploid wheat and tetraploid wheat, which enable the rapid identification of mutation of the gene of interest *in silico* [64]. With the use of a 2014 IWGSC (International Wheat Genome Sequencing Consortium) reference genome, the mutations were identified and these mutations were analyzed using the RefSeqv1.0 genome which is available on the Ensembl Plants website [65].

Moreover, in durum wheat, this approach has been successfully applied to enhance the provitamin A content and it may be applied to increase the iron and zinc biofortification, if suitable genes are present [66]. And some mutagenesis population has been identified for iron and zinc biofortification, which is possible in nature in a limited/

S. No.	Trait	QTL	Mapping population	Chromosome No.	LOD	PVE (%)	Additive effect	References
1	Zn	QGZn.ada	127 RIL	1B	3.7	12	5.07	[68]
2		QGZn.ada		1D	4.2	31	−4.2	
3		QGZn.ada		3A	3.8	14	−2.81	
4		QGZn.ada		6B	7.8	27	−3.54	
5		QGZn.ada		7A	3.1	15	−2.48	
6		QGZn.ada		7B	6.6	25	13.1	
7	Fe	QGFe.ada		2B	5.9	17	3.17	
8		QGFe.ada		2B	5.3	17	2.9	
9		QGFe.ada		6B	3.9	14	−2.14	
10		QGFe.ada		7B	5.6	18	5.78	
11	Zn	QGZn.sar	127 RIL	1B	2.4	9.1	3.1	[69]
12		QGZn.sar		6B	3.1	11.7	−9.8	
13		QGZn.sar		1B	2.5	10.6	4.8	
14		QZneff.sar		6A	2.1	5.9	−4.7	
15		Qzneff.sar		6B	2.4	9.1	−5.1	
16		QshootZn.		1B	3.5	14.6	−4.7	

S. No.	Trait	QTL	Mapping population	Chromosome No.	LOD	PVE (%)	Additive effect	References
17	Fe	QGFe.sar		1B	2.8	10	−4.1	
18		QGFe.sar		3A/3B	3.5	12.1	−3.2	
19		QGFe.sar		5B	4	14.9	2.7	
20		QGFe.sar		5B	4.7	16.9	4	
21	Zn	QGZn.co	200 RIL	6B	3.56	4.2	1.38	[51]
22		QGZn.co		7A	5.47	7.83	1.81	
23		QGZn.co		5A	2.69	14.22	1.73	
24	Fe	QGFe.co		2A	4.12	3	2.77	
25		QGFe.co-		3B	3.65	14.56	−1.71	
26		QGFe.co		4B	5.33	4.41	0.93	
27	Zn	QZn.bhu	138 Double haploid	1B	2.9	11.8	0.4	[70]
28		QZn.bhu		2B	7.7	24	0.5	
29	Fe	QFe.bhu		2B	5	29.6	0.2	
30	Zn	QGzncpk. cimmyt	330 diverse panel	1BS	5.6	10	−2.16	[71]
31		QGzncpk. cimmyt		2 BC	4.9	9	1.96	
32		QGzncpk. cimmyt		3AL	7.9	15	−2.43	
33		QGzncpk. cimmyt		4AS	3.4	7	−1.70	
34		QGzncpk. cimmyt		5BL	3.4*	8	−1.49	
35		QGzncpk. cimmyt		2Bc	6.6	11	2.52	
36		QGzncpk. cimmyt		2D	5.2	26	3.35	
37		QGzncpk. cimmyt		3AL	4.7	11	−2.34	
38		QGzncpk. cimmyt		6AL	4	7	−2.16	
39		QGzncpk. cimmyt		1BS	7	11	−2.47	
40		QGzncpk. cimmyt		2Bc	6.6	10	2.09	
41		QGzncpk. cimmyt		3AL	9	15	−2.56	
42		QGzncpk. cimmyt		4AS	3.1	5	−1.52	
43	Zn	QZn.Y12-13	140 RIL	4BS	4.04	11.7	1.33	[72]
44		QZn.Y13-14_		4BS	6.97	19.6	3.47	
45		QZn.Y13-14		6BL	3.29	9	2.23	
46		QZn.Across		4BS	6.64	17.3	2.7	
47	Fe	QFe.Y12-13		6DS	3.16	9.1	−0.53	
48		QFe.Y13-14		4BS	5.26	12	0.94	
49		QFe.Across		2DS	3.64	7.8	1	
50		QFe.Across		4BS	5.08	10.7	1.03	
51		QFe.Across		6AL	4.18	9.6	0.93	
52		QFe.Across		7DS	6.58	14.5	1.14	

S. No.	Trait	QTL	Mapping population	Chromosome No.	LOD	PVE (%)	Additive effect	References
53	Fe	QGFe.iari	286 RIL	2A	4.1	6.8		[73]
54	Zn	QGZn.iari		2A	13.5	11.1		
55		QGZn.iari		2A	11.8	14.4		
56		QGZn.iari		2A	6.5	8.5		
57	Fe	QFe.shu	118 RIL	2A	3.91	12.58	5.84	[74]
58		QFe.shu		2A	3.71	17.04	6.63	
59		QFe.shu		3D	2.76	44.71	−10.64	
60		QFe.shu		4D	2.54	44.6	−10.66	
61		QFe.shu		7B	2.52	47	10.92	
62		QFe.shu		7D	2.78	8.94	4.9	
63	Zn	QZn.shu		1A	2.97	50.79	−7.11	
64		QZn.shu		4A	2.67	40.22	6.28	
65	Zn	QGzncpk.cimmyt	177 RIL	1BS	5.6	10	−2.16	[75]
66		QGzncpk.cimmyt		2Bc	4.9	9	1.96	
67		QGzncpk.cimmyt		3AL	7.9	15	−2.43	
68		QGzncpk.cimmy		4AS	3.4	7	−1.7	
69		QGzncpk.cimmyt		5BL	3.4	8	−1.49	
70		QGzncpk.cimmy		2Bc	6.6	11	2.52	
71		QGzncpk.cimmyt		2Dd	5.2			
72		QGzncpk.cimmyt		3AL	4.7	11	−2.34	
73		QGzncpk.cimmyt		6AL	4	7	−2.16	
74		QGzncpk.cimmyt		1BS	7	11	−2.47	
75		QGzncpk.cimmyt		2Bc	6.6	10	2.09	
76		QGzncpk.cimmyt		3AL	9	15	−2.56	
77	Zn	QZn.bhu-	185 RIL	2A	3.18	6.68	1.31	[76]
78		QZn.bhu		2B	3.79	6.85	1.11	
79		QZn.bhu		2B	3.6	6.93	1.47	
80		QZn.bhu		2B	2.73	9.3	1.65	
81		QZn.bhu		2B	3.57	6.89	1.26	
82		QZn.bhu		2B	3.58	10.76	1.6	
83		QZn.bhu		2B	4.77	15.69	1.74	
84		QZn.bhu		2B	4.36	16.02	2.1	
85		QZn.bhu		3D	3.63	6.3	1.41	
86		QZn.bhu		3D	4.37	7.18	1.29	
87		QZn.bhu		6A	3.84	7.1	1.14	
88		QZn.bhu		6A	3.82	6.5	1.43	

S. No.	Trait	QTL	Mapping population	Chromosome No.	LOD	PVE (%)	Additive effect	References
89		QZn.bhu		6A	4.55	7.33	1.3	
90		QZn.bhu		6A	4.02	9.47	1.48	
91		QZn.bhu		6A	2.68	6.77	1.12	
92		QZn.bhu		6B	4.35	13.15	2.11	
93	Fe	QFe.bhu		1A	2.97	5.67	0.79	
94		QFe.bhu		1A	3.4	5.63	0.96	
95		QFe.bhu		1A	3.41	5.67	0.69	
96		QFe.bhu		1A	3.71	6.12	0.97	
97		QFe.bhu		1A	3.47	7.23	0.49	
98		QFe.bhu		1A	3.91	8.05	0.84	
99		QFe.bhu		1A	4.06	7.84	0.9	
100		QFe.bhu		1A	4.59	7.71	1.09	
101		QFe.bhu		1A	4.56	7.69	0.78	
102		QFe.bhu		1A	4.37	7.26	1.03	
103		QFe.bhu		1A	6.68	14.56	0.75	
104		QFe.bhu		1A	6.98	15.07	1.24	
105		QFe.bhu		1A	5.69	11.56	1.13	
106		QFe.bhu		1A	9.53	19.94	1.55	
107		QFe.bhu		1A	9.44	17.03	1.76	
108		QFe.bhu		1A	9.37	16.99	1.26	
109		QFe.bhu		1A	9.72	17.51	1.73	
110		QFe.bhu		3B	3.45	6.97	0.85	
111		QFe.bhu		3B	13.96	27.1	2.14	
112		QFe.bhu		3B	13.82	26.87	1.53	
113		QFe.bhu		3B	13.98	26.76	2.07	
114	Zn	QGZn.cimmyt	188 RIL	1B	8.3	15.1	0.531	[72]
115		QGZn.cimmyt		6A	7.91	9.71	0.457	
116		QGZn.cimmy		7B	7.12	16.75	0.424	
117		QGZn.cimmyt		7B	5.42	2.86	0.2	
118		QGFe.cimmyt		3A	7.13	10.35	−0.139	
119		QGFe.cimmyt		4B	4.56	6.69	−0.119	
120		QGFe.cimmyt		5B	4.71	5.49	−0.119	
121		QGZn.cimmyt		1A	8.9	10.78	0.843	
122		QGZn.cimmyt		1B	8.58	11.25	0.814	
123		QGZn.cimmyt		3B	5.43	1.01	0.595	
124		QGZn.cimmyt		3B	7.87	10.93	−0.717	
125		QGZn.cimmyt-		3D	5.8	7.49	0.64	

S. No.	Trait	QTL	Mapping population	Chromosome No.	LOD	PVE (%)	Additive effect	References
126		QGZn.cimmyt		4A	5.17	3.82	−0.455	
127		QGZn.cimmyt		5B	6.6	5.05	−0.576	
128		QGZn.cimmyt		6A	11.88	8.53	−0.907	
129		QGZn.cimmyt		7B	20.76	32.79	−1.290	
130		QGZn.cimmyt		7B	7.77	3.3	−0.493	
131		QGZn.cimmyt		7B	7.03	5.4	−0.527	
132		QGZn.cimmyt		7D	5.23	5.81	−0.596	
133	Fe	QGFe.cimmyt		2A	6.36	14.23	0.112	
134		QGFe.cimmyt		2B	4.98	5.79	0.102	
135		QGFe.cimmyt		3B	5.1	5.81	0.097	
136		QGFe.cimmyt		3B	6.52	7.19	−0.083	
137		QGFe.cimmyt		4A	9.65	21.14	−0.161	
138		QGFe.cimmyt		4D	6.45	14.62	−0.109	
139		QGFe.cimmyt		5B	5.38	11.62	−0.097	

Table 1.
Schematic representation of quantitative trait loci (QTL) of biofortified wheat.

restricted manner, and it has also been observed in gamma-irradiated wheat lines [67]. Therefore, this mutagenesis approach may be utilized for increasing zinc and iron content in wheat crop without using a genetically modified (GM) approach (**Table 1**) [42].

Author details

Anand Kumar[1,2] and Prashant Kaushik[3]*

1 School of Agricultural Sciences and Technology, RIMT University, Mandi Gobindgarh, India

2 Department of Genetics and Plant Breeding, Chandra Shekhar Azad University of Agriculture and Technology, Kanpur, India

3 Institute of Conservation and Improvement of Valencian Agrodiversity, Polytechnic University of Valencia, Spain

*Address all correspondence to: prakau@doctor.upv.es

References

[1] Kumar RR, Singh SP, Rai GK, Krishnan V, Berwal M, Goswami S, et al. Iron and zinc at a cross-road: A trade-off between micronutrients and anti-nutritional factors in pearl millet flour for enhancing the bioavailability. Journal of Food Composition and Analysis. 2022; **22**:104591

[2] Pingali PL. Green revolution: Impacts, limits, and the path ahead. Proceedings of the National Academy of Sciences. 2012;**31**:12302-12308

[3] Aggarwal N, Upadhyay P, Tigadi SB. Biofortification to improve nutrition: A review. International Journal of Current Microbiology and Applied Sciences. 2020;**9**:763-779

[4] Xu Y, An D, Li H, Xu H. Breeding wheat for enhanced micronutrients. Canadian Journal of Plant Science. 2011; **91**:231-237

[5] Graham RD, Knez M, Welch RM. How much nutritional iron deficiency in humans globally is due to an underlying zinc deficiency? Advances in Agronomy. 2012;**115**:1-40

[6] Wessells KR, Brown KH. Estimating the global prevalence of zinc deficiency: Results based on zinc availability in national food supplies and the prevalence of stunting. PLoS One. 2012;**29**:50568

[7] Guzmán C, Medina-Larqué AS, Velu G, González-Santoyo H, Singh RP, Huerta-Espino J, et al. Use of wheat genetic resources to develop biofortified wheat with enhanced grain zinc and iron concentrations and desirable processing quality. Journal of Cereal Science. 2014; **60**:617-622

[8] Adhikari S, Kumari J, Jacob SR, Prasad P, Gangwar OP, Lata C, et al. Landraces-potential treasure for sustainable wheat improvement. Genetic Resources and Crop Evolution. 2022;**69**: 499-523

[9] Adhikari S, Kumari J, Jacob SR, Prasad P, Gangwar OP, Lata C, et al. Landraces-potential treasure for sustainable wheat improvement. Genetic Resources and Crop Evolution. 2022;**18**: 1-25

[10] Hammer K. Resolving the Challenge Posed by Agrobiodiversity and Plant Genetic Resources: An Attempt. Kassel: Kassel University Press; 2004

[11] Shiferaw B, Smale M, Braun HJ, Duveiller E, Reynolds M, Muricho G. Crops that feed the world 10. Past successes and future challenges to the role played by wheat in global food security. Food Security. 2013;**5**:291-317

[12] Maulana F, Ayalew H, Anderson JD, Kumssa TT, Huang W, Ma XF. Genome-wide association mapping of seedling heat tolerance in winter wheat. Frontiers in Plant Science. 2018;**4**:1272

[13] Bhatt R, Hossain A, Sharma P. Zinc biofortification as an innovative technology to alleviate the zinc deficiency in human health: A review. Open Agriculture. 2020;**15**:176-187

[14] King CK, Gale SA, Hyne RV, Stauber JL, Simpson SL, Hickey CW. Sensitivities of Australian and New Zealand amphipods to copper and zinc in waters and metal-spiked sediments. Chemosphere. 2006;**63**: 1466-1476

[15] Balyan HS, Gupta PK, Kumar S, Dhariwal R, Jaiswal V, Tyagi S, et al. Genetic improvement of grain protein content and other health-related

constituents of wheat grain. Plant Breeding. 2013;**132**:446-457

[16] Ackland ML, Michalczyk AA. Zinc and infant nutrition. Archives of Biochemistry and Biophysics. 2016;**611**: 51-57

[17] Wintergerst ES, Maggini S, Hornig DH. Immune-enhancing role of vitamin C and zinc and effect on clinical conditions. Annals of Nutrition and Metabolism. 2006;**50**:85-94

[18] Joachimiak MP. Zinc against COVID-19? Symptom surveillance and deficiency risk groups. PLoS Neglected Tropical Diseases. 2021;**15**:0008895

[19] Prasad AS, Bao B. Molecular mechanisms of zinc as a pro-antioxidant mediator: Clinical therapeutic implications. Antioxidants. 2019;**6**:164

[20] Lai C, Shih TP, Ko WC, Tang HJ, Hsueh PR. Severe acute respiratory syndrome coronavirus 2 (SARS-CoV-2) and coronavirus disease-2019 (COVID-19): The epidemic and the challenges. International Journal of Antimicrobial Agents. 2020;**55**:105924

[21] Akhtar T, Ali G, Sheikh N. Immunosuppressant-induced oxidative stress and iron: A paradigm shift from systemic to intrahepatic abnormalities. Oxidative Medicine and Cellular Longevity. 2020;**31**:2020

[22] Casanueva E, Viteri FE, Mares-Galindo M, Meza-Camacho C, Loría A, Schnaas L, et al. Weekly iron as a safe alternative to daily supplementation for nonanemic pregnant women. Archives of Medical Research. 2006;**37**: 674-682

[23] Available from: https://ods.od.nih.gov/factsheets/Iron-Health Professional. 2020

[24] Harder NH, Hieronimus B, Stanhope KL, Shibata NM, Lee V, Nunez MV, et al. Effects of dietary glucose and fructose on copper, iron, and zinc metabolism parameters in humans. Nutrients. 2020;**25**:2581

[25] Cakmak I, Pfeiffer WH, McClafferty B. Biofortification of durum wheat with zinc and iron. Cereal Chemistry. 2010;**87**:10-20

[26] Sushree Shyamli P, Rana S, Suranjika S, Muthamilarasan M, Parida A, Prasad M. Genetic determinants of micronutrient traits in graminaceous crops to combat hidden hunger. Theoretical and Applied Genetics. 2021;**134**:3147-3165

[27] Garcia-Oliveira AL, Chander S, Ortiz R, Menkir A, Gedil M. Genetic basis and breeding perspectives of grain iron and zinc enrichment in cereals. Frontiers in Plant Science. 2018; **9**:937

[28] Bisht A, Kaur S, Sharma S, Bhandawat A, Bhardwaj S, Garg M, et al. Wheat quality improvement for micronutrients. In: Physiological, Molecular, and Genetic Perspectives of Wheat Improvement. New York: Springer; 2021. pp. 43-69

[29] Marques E, Darby HM, Kraft J. Benefits and limitations of non-transgenic micronutrient biofortification approaches. Agronomy. 2021;**11**:464

[30] Acquaah G. Conventional plant breeding principles and techniques. In advances in plant breeding strategies: Breeding. Biotechnology and Molecular Tools. 2015:115-158

[31] Xu Y, Crouch JH. Marker-assisted selection in plant breeding: From

publications to practice. Crop Science. 2008;**48**:391-407

[32] Gopalareddy K, Singh AM, Ahlawat AK, Singh GP, Jaiswal JP. Genotype-environment interaction for grain iron and zinc concentration in recombinant inbred lines of a bread wheat (Triticum aestivum L.) cross. Indian Journal of Genetics and Plant Breeding. 2015;**75**:307-313

[33] Garcia-Oliveira AL, Chander S, Ortiz R, Menkir A, Gedil M. Genetic basis and breeding perspectives of grain iron and zinc enrichment in cereals. Frontiers in Plant Science. 2018;**9**:937

[34] Joppa LR, Du C, Hart GE, Hareland GA. Mapping gene (s) for grain protein in tetraploid wheat (Triticum turgidum L.) using a population of recombinant inbred chromosome lines. Crop Science. 1997; **37**:1586-1589

[35] Distelfeld A, Uauy C, Fahima T, Dubcovsky J. Physical map of the wheat high-grain protein content gene Gpc-B1 and development of a high-throughput molecular marker. New Phytologist. 2006;**169**:753-763

[36] Uauy C, Brevis JC, Dubcovsky J. The high grain protein content gene Gpc-B1 accelerates senescence and has pleiotropic effects on protein content in wheat. Journal of Experimental Botany. 2006;**1**:2785-2794

[37] Shahzad Z, Rouached H, Rakha A. Combating mineral malnutrition through iron and zinc biofortification of cereals. Comprehensive Reviews in Food Science and Food Safety. 2014;**13**: 329-346

[38] Suenaga K, Singh RP, Huerta-Espino J, William HM. Microsatellite markers for genes Lr34/Yr18 and other quantitative trait loci for leaf rust and stripe rust resistance in bread wheat. Phytopathology. 2003;**93**:881-890

[39] Sab S, Lokesha R, Mannur DM, Jadhav K, Mallikarjuna BP, Yeri S, et al. Genome-wide SNP discovery and mapping QTLs for seed iron and zinc concentrations in chickpea (Cicer arietinum L.). Frontiers. Nutrition. 2020;**7**:559120

[40] Genç Y, Verbyla AP, Torun AA, Cakmak I, Willsmore K, Wallwork H, et al. Quantitative trait loci analysis of zinc efficiency and grain zinc concentration in wheat using whole genome average interval mapping. Plant and Soil. 2009;**314**:49-66

[41] Dekkers JC, Hospital F. The use of molecular genetics in the improvement of agricultural populations. Nature Reviews Genetics. 2002;**3**:22-32

[42] Ali MW, Borrill P. Applying genomic resources to accelerate wheat biofortification. Heredity. 2020;**125**: 386-395

[43] Appels R, Eversole K, Stein N, Feuillet C, Keller B, Rogers J, et al. Shifting the limits in wheat research and breeding using a fully annotated reference genome. Science International Wheat Genome Sequencing Consortium (IWGSC). 2018;**17**:7191

[44] Borrill P, Connorton JM, Balk J, Miller AJ, Sanders D, Uauy C. Biofortification of wheat grain with iron and zinc: Integrating novel genomic resources and knowledge from model crops. Frontiers in Plant Science. 2014; **21**:53

[45] Gaikwad KB, Rani S, Kumar M, Gupta V, Babu PH, Bainsla NK, et al. Enhancing the nutritional quality of major food crops through conventional

and genomics-assisted breeding. Frontiers in Nutrition. 2020;7:533453

[46] Brachi B, Morris GP, Borevitz JO. Genome-wide association studies in plants: The missing heritability is in the field. Genome Biology. 2011;**12**:1-8

[47] Cockram J, Mackay I. Genetic mapping populations for conducting high-resolution trait mapping in plants. Plant genetics and Molecular Biology. 2018:109-138

[48] Bhatta M, Baenziger PS, Waters BM, Poudel R, Belamkar V, Poland J, et al. Genome-wide association study reveals novel genomic regions associated with 10 grain minerals in synthetic hexaploid wheat. International Journal of Molecular Sciences. 2018;**19**:3237

[49] Alomari DZ, Eggert K, Von Wiren N, Alqudah AM, Polley A, Plieske J, et al. Identifying candidate genes for enhancing grain Zn concentration in wheat. Frontiers in Plant Science. 2018;**10**:1313

[50] Zhou Z, Shi X, Zhao G, Qin M, Ibba MI, Wang Y, et al. Identification of novel genomic regions and superior alleles associated with Zn accumulation in wheat using a genome-wide association analysis method. International Journal of Molecular Sciences. 2020;**11**:1928

[51] Liu J, Wu B, Singh RP, Velu G. QTL mapping for micronutrients concentration and yield component traits in a hexaploid wheat mapping population. Journal of Cereal Science. 2019;**88**:57-64

[52] Voss-Fels KP, Cooper M, Hayes BJ. Accelerating crop genetic gains with genomic selection. Theoretical and Applied Genetics. 2019;**132**: 669-686

[53] Velu G, Crossa J, Singh RP, Hao Y, Dreisigacker S, Perez-Rodriguez P, et al. Genomic prediction for grain zinc and iron concentrations in spring wheat. Theoretical and Applied Genetics. 2016; **129**:1595-1605

[54] Velu G, Singh RP, Crespo-Herrera L, Juliana P, Dreisigacker S, Valluru R, et al. Genetic dissection of grain zinc concentration in spring wheat for mainstreaming biofortification in CIMMYT wheat breeding. Scientific Reports. 2018;**10**:1

[55] Nakano Y, Kobayashi Y. Genome-wide association studies of agronomic traits consisting of field-and molecular-based phenotypes. Reviews in Agricultural Science. 2020;**8**:28-45

[56] Hayta S, Smedley MA, Demir SU, Blundell R, Hinchliffe A, Atkinson N, et al. An efficient and reproducible agrobacterium-mediated transformation method for hexaploid wheat (Triticum aestivum L.). Plant Methods. 2019;**15**:1-5

[57] Borrill P. Blurring the boundaries between cereal crops and model plants. New Phytologist. 2020;**228**:1721-1727

[58] Mushtaq M, Mukhtar S, Sakina A, Dar AA, Bhat R, Deshmukh R, et al. Tweaking genome-editing approaches for virus interference in crop plants. Plant Physiology and Biochemistry. 2020;**147**:242-250

[59] Vats S, Kumawat S, Kumar V, Patil GB, Joshi T, Sonah H, et al. Genome editing in plants: Exploration of technological advancements and challenges. Cell. 2019;**4**:1386

[60] Zhang Y, Liang Z, Zong Y, Wang Y, Liu J, Chen K, et al. Efficient and transgene-free genome editing in wheat through transient expression of CRISPR/

Cas9 DNA or RNA. Nature Communications. 2016;7:1-8

[61] Jaganathan D, Ramasamy K, Sellamuthu G, Jayabalan S, Venkataraman G. CRISPR for crop improvement: An update review. Frontiers in Plant Science. 2018;**17**(9): 985

[62] Chen K, Gao C. Targeted genome modification technologies and their applications in crop improvements. Plant Cell Reports. 2014;**33**:575-583

[63] Kiran K. Advanced approaches for biofortification. In: Advances in Agri-Food Biotechnology. New York: Springer; 2020. pp. 29-55

[64] Ali MW, Borrill P. Applying genomic resources to accelerate wheat biofortification. Heredity. 2020;**125**: 386-395

[65] Howe SJ, Mansour MR, Schwarzwaelder K, Bartholomae C, Hubank M, Kempski H, et al. Insertional mutagenesis combined with acquired somatic mutations causes leukemogenesis following gene therapy of SCID-X1 patients. The Journal of Clinical Investigation. 2008;**118**:13-19

[66] Sestili F, Garcia-Molina MD, Gambacorta G, Beleggia R, Botticella E, De Vita P, et al. Provitamin a biofortification of durum wheat through a TILLING approach. International Journal of Molecular Sciences. 2019;**14**: 5703

[67] Kenzhebayeva S, Abekova A, Atabayeva S, Yernazarova G, Omirbekova N, Zhang G, et al. Mutant lines of spring wheat with increased iron, zinc, and micronutrients in grains and enhanced bioavailability for human health. BioMed Research International. 2019;**14**:112-123

[68] Velu G, Singh RP, Cardenas ME, Wu B, Guzman C, Ortiz-Monasterio I. Characterization of grain protein content gene (GPC-B1) introgression lines and its potential use in breeding for enhanced grain zinc and iron concentration in spring wheat. Acta Physiologiae Plantarum. 2017;**39**:1-9

[69] Velu G, Singh RP, Huerta J, Guzmán C. Genetic impact of Rht dwarfing genes on grain micronutrients concentration in wheat. Field Crops Research. 2017;**214**:373-377

[70] Tiwari C, Wallwork H, Arun B, Mishra VK, Velu G, Stangoulis J, et al. Molecular mapping of quantitative trait loci for zinc, iron and protein content in the grains of hexaploid wheat. Euphytica. 2016;**207**:563-570

[71] Velu G, Singh RP. Genomic approaches for biofortification of grain zinc and iron in wheat. In: Quality Breeding in Field Crops. New York: Springer; 2019. pp. 193-198

[72] Crespo-Herrera LA, Govindan V, Stangoulis J, Hao Y, Singh RP. QTL mapping of grain Zn and Fe concentrations in two hexaploid wheat RIL populations with ample transgressive segregation. Frontiers in Plant Science. 2017;**8**:1800

[73] Krishnappa G, Singh AM, Chaudhary S, Ahlawat AK, Singh SK, Shukla RB, et al. Molecular mapping of the grain iron and zinc concentration, protein content and thousand kernel weight in wheat (Triticum aestivum L.). PLoS One. 2017;**12**:0174972

[74] Roshanzamir H, Kordenaeej A, Bostani A. Mapping QTLs related to Zn and Fe concentrations in bread wheat (Triticum aestivum) grain using microsatellite markers. Iranian Journal of Genetics and Plant Breeding. 2013;**2**:10-17

[75] Hao Y, Velu G, Peña RJ, Singh S, Singh RP. Genetic loci associated with high grain zinc concentration and pleiotropic effect on kernel weight in wheat (Triticum aestivum L.). Molecular Breeding. 2014;**4**:1893-1902

[76] Srinivasa J, Arun B, Mishra VK, Singh GP, Velu G, Babu R, et al. Zinc and iron concentration QTL mapped in a Triticum spelta× T. aestivum cross. Theoretical and Applied Genetics. 2014; **127**:1643-1651

Chapter 4

Mitigating Global Food and Nutritional Insecurity: Role of Indigenous Crops

Paul Kweku Tandoh, Irene Akua Idun and Bridget Yayra Bemanu

Abstract

Indigenous plants adapt well to marginal conditions, a situation that is essential for resilient agriculture and sustainable food systems in a rapidly changing global climate. These plants are known to be highly nutritious containing a wide array of antioxidants from their various parts including the leaves, stems, roots, branches, flowers etc. Polyphenols, flavonoids, isoflavonoids are major examples of these antioxidants which are chiefly found in many indigenous fruits, vegetables, nuts and seeds. The cultivation of these crops creates employment, providing a unique hotspot for biodiversity conservation as well as providing raw materials for most industries in the world. This review has also provided particular uses of these crops and their potential to combat food insecurity. The rapid exploitation of these crops and the lack of knowledge on their conservation is a major problem to global nutritional and food insecurity. Additionally, the rapid growth in population and technology will significantly impact the productive uses of these vital species. It is therefore crucial to provide a comprehensive review on the role of some of these plants in combating the food security issues globally.

Keywords: germplasm, malnutrition, antioxidants, fruits and vegetables, nutritional insecurity, global food

1. Introduction

It is an undeniable fact that the consumption of quality food is crucial for the continued survival of living things on this planet. The basic part of the existence of man has been food. Through the ages, we have amassed a lot of knowledge on the use of food to promote healthy growth in children and youth, to maintain good health throughout life, to satisfy unique needs of pregnancy and lactation, and to utilize it to heal from sickness [1]. Food is that which nourishes the body and may also be defined as anything eaten or drunk, which meets the needs for energy, building, regulation and protection of the body. Consuming the proper foods in the right proportions can promote good nutrition and health, which may show in our physical appearance, productivity and emotional well-being. Food functioning in the body is referred to as nutrition, and this definition encompasses everything that occurs to food from the time it is consumed until it is used for various bodily activities. The body requires nutrients in sufficient quantities in order to grow, reproduce, and live a normal, healthy life. Water, proteins, lipids, carbohydrates, minerals, and vitamins are the basic building blocks or constituents of nutrients. There are several nutrients in each of the groups: proteins, fats, carbohydrates, minerals and vitamins; hence the plural form of these words has been used.

2. The concept of food security

The international communiity has made a lot of efforts to define the over all idea of food security and how it has evolved. One of the most basic shifts has been the change from an initial idea where food security was known to be similar to the reliable availability of food towards the contemporary notion in which food is one of the elements of a complex social context that determines livelihoods. The issue of food security is largely determined by this social context and the relative power dynamics among the various interest groups that make up it [2]. "Food security, at the individual, household, national, regional, and global levels [is achieved] when all people, at all times, have physical and economic access to sufficient, safe, and nutritious food to meet their dietary needs and food preferences for an active and healthy life," is the consensus that has emerged from the global debate [3]. Thus, physical availability of food, economic access to food, and proper food utilization—which is a result of the body's capacity to process and use nutrients—are the three main components of the household food security architecture [4–6].

Availability in sufficient quantity of food of an appropriate nature and quality in all parts of the national regions, regardless of its origin (local production, imports or food aid).

Access by all people to the resources needed to obtain the food needed by them for a nutritionally adequate diet. Examples of such resiurces are financial resources, rights of access to the resources required to produce food or to receive it from others.

Stability of access to food, i.e. the assurance of access by people to food even in the midst of natural or economic instabilities.

3. State of global food and nutritional security

FAO [7] reported that, the prevalence of malnutrition in the world is 10.8% and 11.0%, that is, 794 and 815 million people according to the results of 2015 and 2016, respectively. This means that, despite all of our combined efforts, there are more hungry and malnourished individuals than ever before. The World Bank said in 2017 that 45 nations were home to 83 million hungry people. Undernourishment affects no more than 5% of the population in industrialized regions, 13% of the population in developing regions, 20% of the population in African countries, and 13% of the population in Asian countries. Food security has clearly declined in a number of nations in Africa, Southeast Asia, and West Asia. Furthermore, the FAO [8], reported that, under the threat of the COVID-19 pandemic, the number of individuals who experience hunger worldwide grew further in 2020.The Prevalence of Undernourishment (PoU) increased from 8.4 percent to roughly 9.9 percent between 2019 and 2020 after remaining essentially stable from 2014 to 2019, making it more difficult to achieve the Zero Hunger objective in 2030.The 2020 projections ranges from 9.2 to 10.4 percent, based on the assumptions made to reflect the uncertainties around the assessment. As per population, it is predicted that between 720 and 811 million people worldwide would face hunger in 2020. With estimates ranging from 70 to 161 million, the mean of the projected range (768 million) indicates that 118 million more individuals experienced hunger in 2020 than in 2019. The statistics demonstrate pervasive and unsettling geographical disparities. In 2020, 21% of Africans (about one in five of the continent's inhabitants) experienced hunger, which is more than twice the rate in any other region. 3 percentage points have been added to this during the course of a year. This is followed by Latin America and the Caribbean (9.1 percent) and Asia (9.0 percent), with increases of 2.0 and 1.1 percentage points, respectively, between 2019 and 2020. More than half (418 million) of the world's undernourished (768 million) people live in Asia, and more than one-third (282 million) reside in Africa, with Latin America and the Caribbean accounting for around 8% of the total (60 million). In 2020, there were over 14 million more individuals in Latin America and the Caribbean, about 57 million more people in Asia, and 46 million more people in Africa who experienced hunger than in 2019. According to the Food Insecurity Experience Scale, there has been a gradual increase in the percentage of people worldwide who experience moderate or severe food insecurity, from 22.6 percent in 2014 to 26.6 percent in 2019. Then, it increased to 30.4 percent in 2020, the year the COVID-19 pandemic broke out worldwide, nearly tripling its prior five-year average. With a rise of 320 million people in only one year, from 2.05 to 2.37 billion. Thus, roughly one in three people worldwide did not have access to enough food in 2020: Of those, about 928 million (11.9 percent of the world's population) or close to 40 percent experienced extreme food insecurity. In 2020 compared to 2019, there were over 148 million more individuals who were severely food insecure [8]. The regions with the largest increases in moderate or severe food insecurity between 2019 and 2020 were Latin America and the Caribbean (9 percentage points) and Africa (5.4 percentage points), whereas Asia only saw a 3.1-point increase. The prevalence of food insecurity increased for the first time since the start of the Food Insecurity Experience Scale (FIES) data collecting in 2014, even in Northern America and Europe, where the lowest rates of food insecurity are observed. In the year of the COVID-19 pandemic, the gender disparity in the incidence of moderate or severe food insecurity has widened even more globally, with the prevalence of such insecurity among women being 10% greater than that of men in 2020 as opposed to just 6% in 2019. To better understand the relationship between these crucial food access factors and the trends in the various forms of malnutrition, it is useful to track the price of healthy diets and the number of people who cannot afford them. An estimated 3 billion people were unable to afford a nutritious diet in 2019 due to the high cost of healthy meals and persistently high levels of income inequality. The majority of these individuals reside in Asia (1.85 billion) and Africa (1.0 billion), although millions of people in Latin America and the Caribbean (113 million), Northern America, and Europe also lack access to a healthy food (17.3 million) [8].

4. Major causes of food insecurity

The causes of recent increases in hunger and the slowing down of efforts to reduce all forms of malnutrition include conflict, climate variability and extremes, and economic slowdowns and downturns (which are currently being compounded by the COVID-19 epidemic). High and enduring levels of inequality make it harder to mitigate their negative effects. Additionally, millions of people worldwide experience food insecurity and various forms of malnutrition because they are unable to pay for a balanced diet. These major drivers are distinct but not mutually exclusive, as they interact to the detriment of food security and nutrition by creating multiple, compounding impacts at many different points within our food systems.

Household food insecurity (HFI) is as a result of poverty, poor health of household member(s), as well as substandard livelihood and inadequate household management strategies [9]. Nutritional and health security are closely related to but not the same as food security. When the body tissues are exposed to enough amounts of nutrients and other essential substances, people can achieve nutrition security. Health care access security, access to other fundamental human needs, such as sufficient sanitation, and household food security together contribute to nutrition security. Food security and the other determinants of nutrition security are linked with each other [9].

For instance, a family with little financial means might decide not to seek medical care for a child or to not purchase prescribed medications. For food security to be a reality, households need to have unrestricted access to a healthy and nutritious diet. The availability of food in the nation, region, and communities where the homes are located, as well as enough financial means, are prerequisites for access to a nutritious diet The amount of food that is readily available on a national scale depends on how much food is produced domestically, imported, and exported, as well as how much food is wasted or fed to animals. Therefore, ensuring household food security and nutrition security globally depends on maintaining a cheap and sustainable nutritious food supply. Given that the Sustainable Development Goals (SDGs) of the United Nations call for eradicating hunger, achieving food security and improved nutrition, and promoting sustainable agriculture globally, it is crucial to understand and address issues relating to household food security, such as climate change, agricultural commodity price regulations, armed conflicts, and ultimately the health of our planet [10].

5. Consequences of household food insecurity

HFI is a significant biological and psychosocial stressor that, through several pathways, may put people at risk for poor mental, social, and emotional development throughout their lives. The potential connections between HFI, inadequate dietary intakes, nutritional status, and general wellbeing are elements of a biological process. An example of this is a recent study from the United States that details the extremely poor nutritional quality of people with low incomes who are at risk of food insecurity [11]. They consumed extremely little fish, fruit, whole grains, and vegetables in their diets. Indeed, there is a substantial correlation between this dietary pattern and a higher risk of obesity, the metabolic syndrome, chronic diseases including diabetes, and early death [12]. Among those who experience food insecurity, a psycho-emotional pathway includes worry/anxiety, a sense of isolation, deprivation and alienation, distress, and unfavorable family and social interactions [13].

5.1 Poor child development

HFI is a potent stressor that affects children's psycho-emotional, social, behavioral, and intellectual growth in both direct and indirect ways, including problem internalization (such as depression) and externalization [13]. Quantitative research has demonstrated that HFI has an impact on a child's development in addition to the independent impacts of the conventional markers of poverty, such as household income and parental education [13]. The impact of HFI on child development is likely to be influenced by nutritional indicators as well as by psycho-emotional factors affecting how the family functions. Qualitative studies have demonstrated that HFI causes powerful psycho-emotional reactions, such as anxiety, sadness, or anger [14].

5.2 Infectious illnesses

In nations as diverse as Brazil and Haiti, HFI has been linked to an increased risk of childhood malaria, diarrhea, upper respiratory infections, and hospitalizations due to serious infectious diseases. This association may be due to inflammation and a weakened immune system [15, 16].

5.3 Stunting, obesity and chronic diseases

In Brazil, Mexico, and other Latin American nations and areas of the world, HFI has been linked to the double burden of malnutrition, or the coexistence of stunted children and obese mothers living in the same household [17, 18]. Additionally, in Mexico, Ecuador, and other nations, HFI has been linked to significant non-communicable disorders such type 2 diabetes and hypertension [19, 20] which is likely to be related to low dietary quality [21] and stress [19] resulting from HFI.

5.4 Poor mental health among children, youth and adults

In the USA, HFI has been linked to depression and suicidal thoughts in young people, and it has been established that HFI is a major cause of maternal depression worldwide. In turn, maternal depression is a significant risk factor for the poor psychosocial, emotional, and behavioral development of children [22–25].

5.5 Suboptimal sleep patterns

Physical and mental health are significantly impacted by lack of sleep. Among Mexican adults and low-income Latinos with type 2 diabetes in the USA, HFI has been linked to less-than-ideal sleep habits, and this relationship is influenced by stress and anxiety [26].

5.6 Social disruption

The widespread riots and social unrest brought on by the economic crisis and significant food inflation experienced in 2008 have been cited as evidence that HFI is a significant global cause of social unrest and internal struggle [27]. In fact, food scarcities cause such societal disruption that they are viewed as a serious threat to the disintegration of states.

5.7 Environmental sustainability

According to the recent Lancet Series on Planetary Health [28]. The stability of the world's food systems is known to be seriously threatened by the deterioration and destruction of natural ecosystems, which in turn threatens crop diversity. Particularly, climate change has been identified as a key factor in the global destruction or harm to ecosystems. Therefore, preserving environmental sustainability is a major problem for both planetary and human health, making it essential to act right away. It is critical for citizens to be informed about the significance of individual lifestyle choices on the future sustainability and food security of the world since customers drive demand, which in turn drives industry decisions [21]. Similarly, it is critical for governments to develop sustainability policies that create the framework for which consumers can make these choices.

6. The sustainable development goals and food security indicators

According to the United Nation's Sustainable Development Goals, it envisaged that by 2030, end hunger and ensure access by all people, in particular the poor and people in vulnerable situations, including infants, to safe, nutritious and sufficient food all year round. Based on the underlying idea of equitable and sustainable economic growth, the SDGs are a set of 169 objectives and targets, that countries have committed to achieving by 2030. The achievement of all 17 goals from eradicating poverty and hunger to promoting gender justice and planetary sustainability, requires a secure supply of food [10].

7. Food insecurity mitigating strategies-role of indigenous crops

There are a number of mitigation strategies which have been outlined to address the issue of global food and nutritional insecurity of which the cultivation of healthy indigenous crops are included. The World Trade Organization (WTO) stated that "Indigenous agriculture and biological resources are vitally important to the economies, cultures, environment, food security and livelihoods of sub-Saharan Africa [29]. Work done by researchers such as Mbhenyane [30] suggests that there may be great potential for African food systems and its food security if indigenous plants were studied more extensively and included more often as mainstream foods.

With 7000 species used historically as food sources and for a variety of other purposes, indigenous tropical food crops (ITFCs) offer more diversity than alien foods [31, 32]. Given the need for dietary diversity and growing worries about environmental effects like climate change, ITFCs are well-positioned to offer food options that may enhance nutrition, broaden dietary options, and be climate change-adapted [33].

Traditional foods provide a significant chance to diversify the food supply as well as serving as a representation of legacy, identity, and culture. Many cultural identities of many ethnic groups depend on these foods [34, 35]. Therefore, it's crucial to preserve various eating customs, especially those related to food preparation and consumption, as this information might.

Moreover, African Indigenous vegetable and fruit crops are native plants that are utilized for food and medicine. Though these plants have been semi domesticated while others develop as volunteer crops, they have the potential to increase economic development and household nutrition. They are rich in vitamins and minerals with some having high quantities of lysine and essential amino acid which is lacking in fiber and cereal diets.

The focus of this chapter review will be on some indigenous leafy vegetable and fruit crops. Examples of such indigenous leafy vegetables are *Solanum macrocarpon* (African eggplant), Amaranthus spp., *Corchorus olitorius L.* (vegetable jute) and

Hibiscus sabdariffa L. (Roselle). The indigenous fruit crops of interest in this report are *Averrhoa carambola* (star fruit), *Irvingia gabonensis* (wild bush mango), *Annona muricata* L. (soursop), *Synsepalum dulcificum* (miracle berry), *Chrysophyllum albidum* (African star apple) and *Dialium guineense* (African black velvet tamarind). Other crops include; *Vernonia amygdalina* (Bitter leaf), *Solanum torvum* (turkey berries or wild eggplant), *Citrullus colocynthis* (colocynth, bitter apple, wild-gourd, egusi) and *Piper guineense* (African black pepper).

S. macrocarpon (**Figure 1**), originated in West Africa and is cultivated in East and Central Africa where its young stems and fresh leaves are widely consumed [36]. The leaves of African eggplant are considered to be of high nutritional value specifically protein, fat, crude fiber, calcium, and zinc [37, 38]. also reported that, *S. macrocarpon* leaves contain considerable amounts of Sulfur-containing amino acid methionine, polyphenols, especially the flavonoids which is capable of contributing to strong antioxidant properties [36, 39] reported that the leaves consist of 86% water, 6% carbohydrates, 4.6% protein, 1.6% fibers and 1% fat, 14.0% glutamic acid, 13.3% aspartic acid, 7.5% leucine and 6.6% arginine.

Amaranthus (**Figure 2**) is a popular vegetable in Bangladesh, South-East Asia, and Africa. According to [40], the stems and leaves of Amaranthus are a good source of carotenoids, proteins, the essential amino acids methionine and lysine, dietary fiber and minerals such as calcium, potassium, magnesium, copper, phosphorus, iron, zinc and manganese are useful to humans. Amaranthus is also reported to be rich in pigments, such as chlorophylls, carotenoids, amaranthine, betalains, anthocyanin, betacyanin and betaxanthins [41] and natural antioxidant phytochemicals, such as vitamin C, beta carotene, flavonoids, and phenolic acids These natural antioxidant phytochemicals protect the human body against cardiovascular diseases, cancer, arthritis, atherosclerosis, cataracts, retinopathy, emphysema, and neurodegenerative diseases [42–44].

Corchorus olitorius L. (**Figure 3**), is an annual green dicotyledonous leafy vegetable that originated in Africa. It is an important leafy vegetable in the tropics; Egypt, Sudan, India, Bangladesh, and the Caribbean, in tropical Asia such as the Philippines and

Figure 1.
Solanum macrocarpon (African eggplant).

Figure 2.
Amaranthus.

Malaysia, and in North Africa and the Middle East, including Lebanon, Palestine, Syria, Jordon, Tunisia, Turkey, and Cyprus [45]. The leaves of *C. olitorius* L. are used as vegetables. According to [46], edible jute has palatable plant parts rich in lipids, proteins, crude fiber, carbohydrate, vitamins (A, C, E) and the minerals calcium, sodium, potassium, phosphorus and iron. Ramadevi [47] also reported that it contains triterpenes, glucoside, flavonoids, saponins, β-sitosterol, fusidic acid, capsularin and scopoletin.

Hibiscus sabdariffa L. (**Figure 4**), is native to Asia (India to Malaysia) or Tropical Africa and is widely grown in tropics like Caribbean, Central America, India, Africa, Brazil, Australia, Hawaii, Florida and Philippines as a home garden crop [48]. The

Figure 3.
Corchorus olitorius L. (vegetable jute).

Figure 4.
Hibiscus sabdariffa L. (Roselle).

fresh calyces of Roselle are reported to be rich in ascorbic acid, riboflavin, niacin, calcium, carotene and iron [49, 50]. Cisse *et al.* [51], reported that Roselle is rich in amino acids and mineral salts. It was also reported in early studies that Roselle contains protein, carbohydrate, vitamin C, β-carotene and iron [52, 53].

Averrhoa carambola (**Figure 5**), is cultivated extensively in South-East Asia for its fruit [54]. Star fruit is considered a rich source of natural antioxidants and minerals such as magnesium, iron, zinc, manganese, potassium, and phosphorous and can be eaten raw or processed into juices, jams, salads, or pickles [55]. According to [54], star fruit is rich in natural antioxidants such as vitamin C, β-carotene, and

Figure 5.
Averrhoa carambola (star fruit).

Figure 6.
Irvingia gabonensis (wild bush mango).

gallic acid and it contains 60% of cellulose, 27% of hemicelluloses, and 13% of pectin approximately.

Irvingia gabonensis (**Figure 6**), Is native to West and Central Africa where it is cultivated for both its pulp and seeds (nut). Studies on the chemical properties of the fresh seeds identified eighteen (18) amino acids as well as appreciable amounts of nutrients, vitamins and minerals such as calcium, magnesium, potassium, sodium, phosphorus and iron. International Vitamin A Consultative Group (IVACG) reported that, *Irvingia* seeds contain 2.12% monosaturated fatty acids, lauric acid (27.63%), 0.27% polyunsaturated fatty acids and myristic acid (61.68%). The pulp of *Irvingia* contains 81 g water; 15.7 g carbohydrate; 0.9 g protein; 2 g fat; 40 mg phosphorus; 20 mg calcium; 7 mg vitamin C and 2 mg iron. The pulp also contains flavor compounds such as cinnamic acid; zingiberene; dodecanal and dodecanol.

Annona muricata L. (**Figure 7**), is native to the warmest tropical areas in South and North America and is widely distributed throughout tropical and subtropical parts of the world, including India, Malaysia and Nigeria, Australia, Africa [56]. According to [57], soursop contains major minerals such as potassium (K), calcium (Ca), sodium (Na), copper (Cu), iron (Fe) and magnesium (Mg), which can help provide essential nutrients and elements to the human body. Morton *et al.* [58], evaluated the composition of soursop and reported that it contains 53.1–61.3 Calories, 82.8 g Moisture, 1.00 g Fat 0.97 g Protein, 14.63 g Carbohydrates, 0.79 g Fiber, 60 g Ash, 10.3 mg Calcium, 27.7 mg Phosphorus, 0.64 mg Iron, 0 Vitamin A (β-carotene), 0.11 mg Thiamine, 0.05 mg Riboflavin, 1.28 mg Niacin, 29.6 mg Ascorbic acid, 11 mg Tryptophan, 7 mg Methionine and 60 g Lysine. Abbo *et al.* [59] studied the nutrient composition of soursop fruit, he reported that soursop fruit contains 81.9–93.6% carbohydrate (glucose and fructose), 278 mg K, 27 mg P, 14 mg Ca, 3.3 g Fiber and 0.3 g/100 g Fat.

Synsepalum dulcificum (**Figure 8**), was first discovered in West and Central Africa, specifically in Congo, Ghana and Nigeria [60, 61]. The berries are a good source of both essential amino acids (lysine, leucine, isoleucine, phenylalanine, threonine etc.) and non-essential (glycine, proline, serine, tyrosine) amino acids [62]. Miracle fruit contains highly essential antioxidantive phytochemicals like epicatechin, rutin,

Figure 7.
Annona muricata L. (soursop).

Figure 8.
Synsepalum dulcificum (miracle berry).

quercetin, myricetin, kaempferol, gallic, ferulic and syringic acid, delphinidin glucoside, cyanidin galactoside and malvidin galactoside, a-tocotrienol, a- and c-tocopherol and lutein. Its proximate berry has been reported by He *et al.* [61] to include ten different fatty acids in the seed oil with a total unsaturated fatty acid content of 52.7%. Vitamins A, C, D and K were reported to be present. The berry was reported to be high in vitamin C [61, 63]. The fruit pulp produced high amounts of mineral content of Ca (100 ppm), Fe (24.20 ppm), Zn (9.49 ppm), Cu (6.22 ppm), Cr (0.01 ppm) and Co (0.01 ppm) with no lead detected. Nkwocha *et al.* [63] reported a high amount of flavonoids (57.01%) in Miracle berry. Tryptophan (8.06%), histidine (0.4%), isoleucine (0.7%), leucine (0.6%), lysine (0.6%), methionine (1.05%), phenylalanine (0.7%), threonine (1.1%), and valine (0.69%) are the essential and non-essential amino acids that are profiled in this plant. According to Lim [64] the principal compound of the fruit, "miraculin" is made up of sugars (glucosamine, mannose, fructose, xylose, and galactose), Nitrogen, Carbohydrates, and nearly 191 amino acid residues.

Chrysophyllum albidum (**Figure 9**), is originated from the central, eastern and West Africa and is common in both urban and rural centres in Nigeria. It is reported as an excellent source of vitamin C, iron [65]. It is also known to be rich in minerals especially K (potassium) and Mg (magnesium) [66, 67]. According to Okoli an Okere [68], African star apple is a rich source of anti-inflammatory and anti-hemorrhoidal compounds.

Dialium guineense (**Figure 10**), Is found in Central and West African countries such as Cameroon, Central African Republic, Chad, Benin, Burkina Fasso, Ivory Coast, Ghana, the Guineas, Liberia, Mali, Senegal, Sierra Leone, and Togo [69]. It is rich in minerals such as magnesium (Mg), sodium (Na), iron (Fe), potassium (K) and beta-carotene (Vitamin A), copper (Cu), sugars and tartaric acid, citric acid, malic acid, ascorbic acid and Niacin. As anticipated, this fruit also has high levels of anti-oxidant.

Vernonia amygdalina (**Figure 11**), is small shrub used as a leafy vegetable and grows in tropical Africa. It is reported to be have been used to alleviate micronutrients malnutrition because the leaves are rich in vitamins and mineral elements including potassium, iron, phosphorus, calcium, copper, zinc, ascorbic acids and folic acid [70]. The leaves and young shoots of *Vernonia amygdalina* are used as a green vegetable in soups and stews and in Nigeria and in some other African countries [71]. According to Musa *et al.* [72] leaves are a rich source of proteins, a-carotene, vitamin C, iron, phosphorus, magnesium, calcium, copper, sodium, zinc and potassium. (Oyowele *et al.* [73] also reported bitter leaf to contain high amounts of alkaloids and flavonoids hence considered to have anti-inflammatory activity.

Solanum torvum (**Figure 12**) is native to Africa and West Indies [74]. It is distributed in the tropical and subtropical mainly in India, China, Pakistan, Philippines and

Figure 9.
Chrysophyllum albidum (African star apple).

Figure 10.
Dialium guineense (African black velvet tamarind).

Figure 11.
Vernonia amygdalina (bitter leaf).

tropical America [75]. Its fruits and leaves are eaten as vegetables in soups and stews [76]. *S torvum* fruits are reported as a good source of fiber, calories, proteins, vitamins and minerals [77]. It is also reported to be high in calcium, iron, zinc, phosphorus, vitamin C and fatty acids.

Citrullus colocynthis (**Figure 13**), originated in the tropical regions of Asia and Africa. It is a vine plant found in the arid region which grows in sandy soil and is now extensively grown in the Mediteranean region and Sahara-Arabian phyto-geographic region of Africa [78]. In India, it is usually found in sandy lands of Northern West region such as Sind, Punjab, Central region, Southern region and Coromandal coastal area [79]. According to National Research Council [80], it contains high amount of protein (30%), 10% carbohydrates, 4% ash content and 3% of fiber content. It is also reported to contain nutritional compounds and different bioactive compounds

Figure 12.
Solanum torvum (Turkey berries or wild eggplant).

Figure 13.
Citrullus colocynthis (colocynth, bitter apple, wild-gourd, egusi, etc.).

such as alkaloids, essential oils, flavonoids, glycosides etc. C. colocynthis contains a specific bioactive compound named as curcurbitacins (A, B, C, D, E, I, J, K and L) as well as colocynthosides (A and B) respectively [81]. In addition to the high oil content of *C. colocynthis* seeds, it is also a rich source of protein and contain all essential amino acids in appropriate amount which makes the protein quality of its seeds superior and equally important as legumes [80]. *C. colocynthis* seed was considered a potential source of calcium (Ca) and potassium (K) in concentration at 569 mg/100 g and 465 mg/100 g respectively. A report by Zaini et al. [82] stated that *C. colocynthis* seeds contain iron (Fe), zinc (Zn) and phosphorus (P) in abundant amounts as well as considerable amount of calcium (Ca) and niacin.

Figure 14.
Piper guineense (African black pepper).

Piper guineense (**Figure 14**) is native to tropical Western Africa [83]. This is a spice plant with more than 700 species in the tropical and subtropical regions of the world [84] and it is commonly called Ashanti pepper, Benin pepper, Guinea pepper, false cubeb, Uziza in Igbo and Iyere in Yoruba [85]. According to Nwankwo *et al.* [86], the proximate analysis of *P. guineense* shows it contains crude protein, fat, carbohydrate, vitamins and minerals. It is also said to contain considerable amounts of vitamin C [87]. A study conducted on *P. guineense* leaves by Nwankwo *et al.*, [86] revealed that it is high in ash which implies it is high in mineral content: copper (Cu), zinc (Zn), calcium (Ca), magnesium (Mg) and potassium (K) [88]. *P. guineense* contains vitamin A and traces of vitamin B1, B2 and E. According to [89]. the proximate composition of *P. guineense* seeds shows that the seeds contain 6.33 + 0.02% Ash, 8.79 + 0.01% Crude fiber, 9.89 + 0.07% Crude fat and 5.86 + 0.04% Crude protein. There are appreciable amounts of essential minerals such as Calcium (Ca), Magnesium (Mg), Potassium (K), Sodium (Na), Phosphorous (P) and Iron (Fe).

8. General uses of indigenous crops in combating food and nutritional insecurity

Many rural communities have access to traditional crops that are rich in micronutrients, which are likely to serve as a long-term strategy to eliminate food insecurity. The diversity of indigenous crops has the potential to augment the nutrient composition of family diets and may contribute to household food security and the alleviation of hidden hunger which is a result of a lack of dietary diversity, usually linked to poor consumption of fruit and vegetables in general. Different nutrients found in food keep the body healthy. It is essential to eat a variety of foods in order to get all the vitamins and minerals required from the diet. Essential nutrient content in food is affected by a number of things, including how it was prepared, the temperature at which it was cooked, how long it took before it was removed from the heat source, etc. Cooking and processing can damage some nutrients and phytochemicals in plant foods. Fruits and vegetables quickly lose their nutritional value.

In order to preserve the nutrients in vegetables, it is therefore advised to avoid overcooking them. Households should employ a variety of methods for preparing vegetables, such as sautéing, steaming, baking, and stir-frying. Vegetables that can be eaten raw can be used in salads. Vegetables should be cooked with very little water and at a low temperature. Soups and stews are excellent ways to retain nutrients that are typically lost in cooking water. Solanum macrocarpon (African eggplant), Amaranthus spp., *Corchorus olitorius* L. (vegetable jute), *Vernonia amygdalina* (Bitter leaf), *Solanum torvum* (turkey berries or wild eggplant), *Citrullus colocynthis* (colocynth, bitter apple, wild-gourd, egusi) and *P. guineense* (African black pepper) are all indigenous vegetables used in the preparation of soups and stews where they are found in West Africa. For continuous availability of these vegetables in their fresh state, folks are encouraged to go into the cultivation of these vegetables to reduce the travel and storage time, buy fresh local produce, shop often and eat produce soon after it has been purchased.

Fruits should either be consumed unprocessed or transformed into juice and concentrates. Indigenous fruits such as *Averrhoa carambola* (star fruit), *Irvingia gabonensis* (wild bush mango), *Annona muricata* L. (soursop), *Synsepalum dulcificum* (miracle berry), *Chrysophyllum albidum* (African star apple) and *Dialium guineense* (African black velvet tamarind) can be consumed fresh or made into juice or smoothies. This will create employment for the youth and contribute to sustainable development. Increased consumption of these indigenous crops will undoubtedly lead to greater demand, increased production and creation of markets and will thus enhance rural economies.

9. Conclusion and wayforward

Without a doubt this review has elucidated information on the concept of food security and the empirical evidence on the current state of the global food and nutritional security whiles outlining the usefulness of some selected indigenous crops which can serve as healthy foods for consumption and raw materials for industrial uses. It has also provided the direct causes and consequences Household Food Insecurity has brought. The COVID-19 pandemic has been a huge setback such that many economies have not been able to recover and thereby making it difficult for the achievement of the hunger eradication agenda envisaged the UNSDGs in 2030. Even though indigenous crops have been known to be of advantage in reducing hunger and malnutrition this review made it clear that further research in this area is imperative, more studies are required to contribute new knowledge to science, hence increasing awareness on traditional uses and management of such crops. Additionally, most African countries, the ethno-botany of wild food resources is poorly documented and patchy, consisting of lists of plant names, providing little to no information on their use and management.

Conflict of interest

The authors declare no conflict of interest.

Author details

Paul Kweku Tandoh*, Irene Akua Idun and Bridget Yayra Bemanu
Department of Horticulture, Kwame Nkrumah University of Science and Technology, Ghana

*Address all correspondence to: pktandoh.canr@knust.edu.gh

References

[1] Mudambi SR. Fundamentals of Foods, Nutrition and Diet Therapy. New Age International; 2007

[2] Carr E. Postmodern conceptualization, modernist applications: Rethinking the role of society in food security. Food Policy. 2006;**31**:14-29

[3] Food and Agriculture Organization (FAO). World Food Summit. Rome, Italy: FAO; 1996

[4] FAO. Regional overview of food insecurity: African food security prospects brighter than ever. In: Food and Agriculture Organization of the United Nations. Accra, Ghana; 2015. pp. 1-37

[5] Ramakrishnan U. Prevalence of micronutrient malnutrition worldwide. Nutrition Reviews. 2002;**60**(5 Pt 2):S46-S52

[6] Smith MD, Rabbitt MP, Coleman-Jensen A. Who are the World's food insecure? New evidence from the food and agriculture Organization's food insecurity experience scale. World Development. 2017;**93**:402-412

[7] Food and Agriculture Organization (FAO). World Food Summit. Rome, Italy: FAO; 2017. Available from: http://www.fao.org/wfs/index_en.htm

[8] Food and Agriculture Organization (FAO). World Food Summit. Rome, Italy: FAO; 2021. Available from: https://www.fao.org/3/cb4474en/online/cb4474en.html

[9] Perez-Escamilla R, Segall-Corrêa AM. Food insecurity measurement and indicators. Revista de Nutrição (Brazil). 2008;**21**:15-26

[10] Food and Agriculture Organization (FAO). The State of Food and Agriculture: Climate Change, Agriculture and Food Security. Rome, Italy: FAO; 2016. Available from: http://www.fao.org/3/a-i6030e.pdf

[11] Leung CW, Ding EL, Catalano PJ, Villamor E, Rimm EB, Willett WC. Dietary intake and dietary quality of low-income adults in the supplemental nutrition assistance program. The American Journal of Clinical Nutrition. 2012;**96**(5):977-988

[12] Millen BE, Abrams S, Adams-Campbell L, Anderson CA, Brenna JT, Campbell WW, et al. The 2015 dietary guidelines advisory committee scientific report: Development and major conclusions. Advances in Nutrition. 2016;**7**(3):438-444

[13] Perez-Escamilla F, de Toledo Vianna RP. Food insecurity and the behavioral and intellectual development of children: A review of the evidence. Journal of Applied Research on Children. 2012;**3**(1):9

[14] Fram MS, Frongillo EA, Jones SJ, Williams RC, Burke MP, DeLoach KP, et al. Children are aware of food insecurity and take responsibility for managing food resources. Journal of Nutrition. 2011;**141**(6):1114-1119

[15] Gubert MB, Spaniol AM, Bortolini GA, Pérez-Escamilla R. Household food insecurity, nutritional status and morbidity in Brazilian children. Public Health Nutrition. 2016;**19**(12):2240-2245

[16] Pérez-Escamilla R, Dessalines M, Finnigan M, Pachón H, Hromi-Fiedler A, Gupta N. Household food insecurity is associated with childhood malaria in rural Haiti. Journal of Nutrition. 2009;**139**(11):2132-2138

[17] Rivera JA, Pedraza LS, Martorell R, Gil A. Introduction to the double burden of undernutrition and excess weight in Latin America. American Journal of Clinical Nutrition. 2014;**100**(6):1613S-1616S

[18] Kulkarni VS, Kulkarni VS, Gaiha R. "double burden of malnutrition": Reexamining the coexistence of undernutrition and overweight among women in India. International Journal of Health Services. 2017;**47**(1):108-133

[19] Laraia BA. Food insecurity and chronic disease. Advances in Nutrition. 2013;**4**(2):203-212

[20] Weigel MM, Armijos RX, Racines M, Cevallos W, Castro NP. Association of Household Food Insecurity with the mental and physical health of low-income urban Ecuadorian women with children. Journal of Environmental and Public Health. 2016

[21] Millen BE, Abrams S, Adams-Campbell L, Anderson CA, Brenna JT, Campbell WW, et al. The 2015 dietary guidelines advisory committee scientific report: Development and major conclusions. Advances in Nutrition. 2016;7(3):438-444

[22] Pérez-Escamilla R, Villalpando S, Shamah-Levy T, Méndez-Gómez HI. Household food insecurity, diabetes and hypertension among Mexican adults: Results from Ensanut 2012. Salud Publica de Mexico. 2014;**56** (Suppl. 1):s62-s70

[23] Perez-Escamilla R. Food Insecurity in Children: Impact on Physical, Psychoemotional and Social Development. In Modern Nutrition in Health and Disease. 11th ed, edited by C. A. Ross, B. Caballero, R. J. Cousins, K. L. Tucker, and T. R. Ziegler. Baltimore, MD: Lippincott Williams & Wilkins, chapter. 2013;**72**:1006-1015

[24] Britto PR, Lye SJ, Proulx K, Yousafzai AK, Matthews SG, Vaivada T, et al. Early childhood development interventions review group, for the lancet early childhood development series steering committee. Nurturing care: Promoting early childhood development. Lancet. 2017;**389**(10064):91-102

[25] Jordan ML, Perez-Escamilla R, Desai MM, Shamah-Levy T. Household food insecurity and sleep patterns among Mexican adults: Results from ENSANUT-2012. Journal of Immigrant and Minority Health. 2016;**18**(5):1093-1103

[26] Bermúdez-Millán A, Pérez-Escamilla R, Segura-Pérez S, Damio G, Chhabra J, Osborn CY, et al. Psychological distress mediates the association between food insecurity and suboptimal sleep quality in Latinos with type 2 diabetes mellitus. Journal of Nutrition. 2016;**146**(10):2051-2057

[27] Brown LR. Could food shortages bring down civilization? Scientific American. 2009;**300**(5):50-57

[28] The Lancet. Safeguarding human health in the Anthropocene epoch: Report of the Rockefeller Foundation. Lancet Commission on Planetary Health. 2015

[29] Akinola R, Pereira LM, Mabhaudhi T, De Bruin FM, Rusch L. A review of indigenous food crops in Africa and the implications for more sustainable and healthy food systems. Sustainability. 2020;**12**(8):3493

[30] Mbhenyane XG. Indigenous foods and their contribution to nutrient requirements. South African Journal of Clinical Nutrition. 2017;**30**(4):5-7. DOI: 10.17660/ActaHorti.2016.1132.16

[31] Jaenicke H, Höschle-Zeledon I. Strategic Framework for Underutilized Plant Species Research and Development: With Special Reference to Asia and the Pacific, and to Sub-Saharan Africa. Rome, Italy: International Centre for Underutilised Crops, Colombo Sri Lanka and Global Facilitation Unit for Underutilised Species; 2008

[32] Bharucha Z, Pretty J. The roles and values of wild foods in agricultural systems. Philosophical Transactions of the Royal Society B: Biological Sciences. 2010;**365**(1554):2913-2926

[33] Pitso FS, Lebese MR. Traditional uses of wild edible plants in arid areas of South Africa. Journal of Human Ecology. 2014;**48**(1):23-31

[34] Sharif MSM, Zahari MSM, Nor NM, Muhammad R. The importance of knowledge transmission and its relation towards the Malay traditional food practice continuity. Procedia-Social and Behavioural Science. 2016;**222**:567-577

[35] Gundo N. Early Childhood Development Stakeholder Knowledge and Perceptions of Indigenous Food in Fetakgomo Municipality, Skhukhune District Limpopo Province. Durban, South Africa: University of KwaZulu-Natal; 2017

[36] Schippers RR. African Indigenous Vegetables. An Overview of the Cultivated Species. UK: University of Greenwich, Natural Resources Institute/ ACP-EU; 2000. p. 214

[37] Oboh G, Ekperigin MM, Kazeem MI. Nutritional and haemolytic properties of eggplants (*Solanum macrocarpon*) leaves. Journal of Food Composition and Analysis. 2005;**18**(2-3):153-160. DOI: 10.1016/j.jfca.2003.12.013

[38] Komlaga G, Sam GH, Dickson RA, Mensah MLK, Fleischer TC. Pharmacognostic studies and antioxidant properties of the leaves of *Solanum macrocarpon*. Journal of Pharmaceutical Sciences and Research. 2014;**6**(1):1-4

[39] Lin JY, Tang CY. Determination of total phenolic and flavonoid contents in selected fruits and vegetables, as well as their stimulatory effects on mouse splenocyte proliferation. Food Chemistry. 2007;**101**(1):140-147

[40] Oba SU. S protein, dietary fiber, minerals, antioxidant pigments and phytochemicals, and antioxidant activity in selected red morph Amaranthus leafy vegetable. PLoS One. 2019a. DOI: 10.1371/journal.pone.0222517

[41] Sarker U, Oba S, Daramy MA. Nutrients, minerals, antioxidant pigments and phytochemicals, and antioxidant capacity of the leaves of stem amaranth. Scientific Reports. 2020;**10**(1):1-9

[42] Stefensen SK. Variations in the polyphenol content of seeds of field grown Amaranthus genotypes. Food Chemistry. 2011;**129**:131-138

[43] Repo-Carrasco-Valencia R, Hellstrom JK, Pihlava JM, Mattila PH. Flavonoids and other phenolic compounds in Andean indigenous grains: Quinoa (Chenopodium quinoa), kaniwa (*Chenopodium pallidicaule*) and kiwicha (*Amaranthus caudatus*). Food Chemistry. 2020;**120**:128-133

[44] Dusgupta N. De B. Antioxidant activity of some leafy vegetables of India: A comparative study. Food Chemistry.2007;**101**:471-474

[45] Musa A, Ezenwa MIS, Oladiran JA, Akanya HO. Ogbadoyi EO effect of soil nitrogen levels on some micronutrients, antinutrients and toxic substances *in*

Corchorus olitorius in Minna, Nigeria. African Journal of Agricultural Research. 2010;**5**(22):3075

[46] Choudhary SB, Sharma HK, Karmakar PG, Kumar AA, Saha AR, Hazra P, et al. Nutritional profile of cultivated and wild jute (Corchorus) species. Australian Journal of Crop Science. 2013;7(13):1973-1982

[47] Ramadevi D. Phytochemical examination of *Corchorus capsularis* leaves. Int. J. Current Pharmaceutical Review Research. 2014;**4**(2):42-46

[48] Gautam RD. Sorrel–a lesser-known source of medicinal soft drink and food in Indian. National Production Radiance. 2004;**3**(5):338-342

[49] Singh P, Khan M, Hailemariam H. Nutritional and health importance of hibiscus sabdariffa: A review and indication for research needs. Journal of Nutritional Health and Food Engineering. 2017;**6**(5):1-4

[50] Mahadevan N, Shivali OE, Kamboj P. *Hibiscus sabdariffa* Linn. An overview. Natural Product Radiance. 2009;**8**:77-83

[51] Cisse M, Domier M, Sakho M, N'Diaye A, Reynes M, Sock O. Le bissap (*Hibiscus sabdariffa* L): Composition etprincipales utilizations. Fruits. 2009;**64**(3):179-185

[52] Singh P, Khan M, Hailemariam H. Nutritional and health importance of *hibiscus sabdariffa*: A review and indication for research needs. Journal of Nutritional Health and Food Engineering. 2017;**6**(5):1-4

[53] Ismail A, Ikram EHK, Nazri HSM. Roselle (*Hibiscus sabdariffa* L.) seeds nutritional composition protein quality and health benefits. Food. 2008;**2**(1):1-16

[54] Muthu N, Lee SY, Phua KK, Bhore SJ. Nutritional, medicinal and toxicological attributes of star-fruits (*Averrhoa carambola* L.): A review. Bioinformation. 2016;**12**:420-424

[55] Moresco HH, Queiroz GS, Pizzolatti MG, Brighente IMC. Chemical constituents and evaluation of the toxic and anti-oxidant activities of Averrhoa carambola leaves. Revista Brasileira De Farmacognosia. 2012;**22**:319-324

[56] Adewole SO, Caxton-Martins EA. Morphological changes and hypoglycemic effects of Annona muricata Linn. (Annonaceae) leaf aqueous extract on pancreatic B-cells of streptozotocin-treated diabetic rats. African Journal of Biomedical Research. 2006;**9**:173-187. Technol. Res. 8 (3):239-243. DOI: 10.3136/fstr.8.239

[57] Gyamfi K, Sarfo D, Nyarko B, Akaho E, Serfor-Armah Y, Ampomah-Amoako E. Assessment of elemental content in the fruit of graviola plant, *Annona muricata,* from some selected communities in Ghana by instrumental neutron activation analysis. Elixir Food Science. 2011;**41**:5671-5675

[58] Morton JF. Fruits of Warm Climates. 1987

[59] Abbo ES, Olurim T, Odeyemi G. African Journal of Biotechnology. 2006;**5**:108-112

[60] Sulaini AA, Sabran SF. Edible and medicinal plants sold at selected local markets in Batu Pahat, Johor, Malaysia. In: Paper Presented at the AIP Conference Proceedings. 2018

[61] He Z, Tan JS, Abbasiliasi S, Lai OM, Tam YJ, Ariff AB. Phytochemicals, nutritionals and antioxidant properties of miracle fruit *Synsepalum dulcificum*.

Indigenous Crop Production. 2016;**86**: 87-94

[62] Njoku NE, Ubbaonu CN, Alagbaoso SO, Eluchie CN, Umelo MC. Amino acid profile and oxidizable vitamin content of *Synsepalum dulcificum* berry (miracle fruit) pulp. Food Science and Nutrition. 2015;**3**(3):252-256

[63] Nkwocha C, Njoku O, Ekwueme F. Proximate and micronutrient analyses of *Synsepalum dulcificum* pulp. Scientific Research. 2014;**2**(1):2201-2796

[64] Lim TK. Edible medicinal and non-medicinal plants, 1. In: Morton LJ, editor. Fruits of Warm Climates, Soursop. Springer; 2012. pp. 75-80

[65] Adisa VA, Fajola A. Postharvest seed rots of three species of citrus in southwestern Nigeria. Indian Phytopathology. 2000;**35**:595-603

[66] Adeleye S, Orji C, Braide WA. kaluka C, Phytochemistry and antimicrobial property of fruits of *Chrysophyllum albidum* against selected clinical isolates. International Letters of Natural Sciences. 2016;**55**:44-51

[67] Iheanyichukwu E, Caleb NJ, Uche OI, Ezkwa AS, Nkechinyere EA. Nutritional properties and antioxidant activity of *Chrysophyllum africanum* leaves and pulp. Journal of Molecular Pharmaceutical Organization Process Research. 2017;**5**:139

[68] Okoli BJO. OS antimicrobial activity of the phytochemical constituents of *Chrysophyllum albidum* G. Don Holl (African star apple). Plant, Journal of Research in National Development. 2010;**8**:1-22

[69] Ogungbenle HN, Ebadan P. Nutritional qualities and amino acid profile of velvet tamarind pulp. British Biomedical Bulletin. 2014;**2**:6-60

[70] Baiyeri PK, Ugese FD, Nwobodo CE. Diversity, food, nutrition and medicinal importance of some selected less-known African indigenous tropical plant species: Advocacy for global research attention. CABI Reviews. 2019:1-12

[71] Masaba SC. The anti-malarial activity of *V. amygdalina* Del. *(Compositae)*. Transactions of the Royal Society of Tropical Medicine and Hygiene. 2000;**94**:64-695

[72] Musa A, Ogbadoyi EO, Oladiran JA, Ezenwa MS, Akanya HO. Effects of reproductive phase on some micronutrients, anti-nutrients and toxic substances in V. amygdalina. African Journal of Plant Science. 2011;**5**(9):525-530

[73] Oyowele BY, Olaleye SB, Elegba RA. Anti-inflammatory and analgesic activities of leaf extract of Landolphiaowarensis. African Journal of Biomedical Research. 2002;**4**(3):131-133

[74] Adjanohoun J, Aboubakar N, Dramane K, Ebot E, Ekpere A, Enoworock G, et al. Traditional medicine and pharmacopeia-contribution to ethnobotanical and floristic studies in Cameroon. In: CNPMS. Porto-Novo, Benin; 1996. pp. 50-52

[75] Nasir JY. Solanaceae. In: Ali SI, Nasir E, editors. Flora of Pakistan, Fascicle 168. Islamabad: Pakistan Agricultural Research Council; 1985. p. 61

[76] Kala CP. Ethnomedicinal botany of the Apatani in the eastern Himalayan region of India. Journal of Ethnobiology and Ethnomedicine. 2005;**1**:1-8

[77] Melila M, Dossou BR, Etse KD, Sika F, Batchazi K, Kanabiya E, et al. Biochemical study and evaluation of the nutritional value of *Solanum*

torvum (Swartz) fruits used as fruiting vegetables in Togo. Journal of Food and Nutrition Research. 2021;**9**(11):579-584

[78] Amamou F, Bouafia M, Chabane-Sari D, Kaouthar MR. *Citrullus colocynthis*: A desert plant native in Algeria, effects of fixed oil on blood homeostasis in Wistar rat. Journal of National Production Plant Resources. 2011;**1**(3):1-7

[79] Borhade P, Deshmukh T, Patil V, Khandelwal K. Review on *Citrullus colocnythis*. International Journal of Research Pharmaceutical Chemistry. 2013;**3**(1):46-53

[80] National Research Council. Lost Crops of Africa. Washington, DC: National Academies Press; 2006 (ISBN 978-0-309-10333-6)

[81] Hussain AI, Rathore HA, Sattar MZA, Chatha SAS, Sarker SD, Gilani AH. *Citrullus colocynthis* (L.) Schrad. (bitter apple fruit): A review of its phytochemistry, pharmacology, traditional uses and nutritional potential. Journal of Ethnopharmacology. 2014;**155**(1):54-66

[82] Zaini NAM, Anwar F, Hamid AA, Saari N. Kundur [Benincasa hispida (Thunb.) Cogn.]: A potential source for valuable nutrients and functional foods. Food Research International. 2011;**44**:2368-2376

[83] Dalziel IM. The useful Plants in West Tropical Africa Handbook. 2nd printing, Crown Agents. 1955

[84] Anyanwu CU, Nwosu GC. Assessment of antimicrobial activity of aqueous and ethanolic extracts of *Piper guineense* leaves. J Med Res. 2012;**48**(10):337-439

[85] Katzer and Gernot. Gernot Katzer Spice pages. 2015

[86] Nwankwo CS, Ebenezer IA, Ikpeama AI, Asuzu FO. The nutritional and anti-nutritional values of two culinary herbs–Uziza leaf (*Piper guineense*) and scent leaf (*Ocimum gratissium*) popularly used in Nigeria. International Journal of Science Engineering Research. 2014;**5**:1160-1163

[87] Okonkwo C, Ogu A. Nutritional evaluation of some selected spices commonly used in south eastern part of Nigeria. Journal of Biology, Agriculture and Healthcare. 2014;**4**(15):45-51

[88] Idris S, Ndamitso MM, Yisa J, Dauda BEN, Jacob JO. The proximate and mineral composition of the leaves and stems of *Balanites aegytiaca*. International Journal of Applied Biological Research. 2011;**2**(1):76-87

[89] Negbenebor CA, Godiya AA, Igene JO. Evaluation of *Clarias anguillains* treated with spice of *P. guineense* for washed mice and Kama book type product, food composition. Analysis. 1999;**2**:12-315

Chapter 5

Bottle Gourd Landraces and Its Potential Contribution to Food Security

Nkanyiso J. Sithole, Khayelihle Ncama, Lembe S. Magwaza, Nozipho M. Motsa and Jacob Mashilo

Abstract

Issues surrounding water scarcity and malnutrition in the rural household of sub-Saharan Africa continue to be a problem and pose threat to food security. In such cases, traditional and underutilized crops have been proposed because of their likely suitability in these marginal areas of crop production characterized by abiotic (drought and heat) and biotic environmental stresses (pest and diseases). Bottle gourd is one such crop with multipurpose use and a huge potential to contribute to food security. The crop is grown for its leaves, fruits, and seeds from landraces providing important nutrition for both humans and livestock. A lot has been documented about its medical properties. The crop exhibits wide genetic variation for qualitative and quantitative traits, which can be used for cultivar development. The objective of this review was to provide information on why bottle gourd is an important food security crop in sub-Saharan Africa. The first section of the paper presents water scarcity, food production, and climate change. This is followed by the section on the neglected underutilized crops species. Then the section of drought tolerance of the crop is presented, and lastly, the section on potential contribution of the crop to food security is presented.

Keywords: calabash, water scarcity, neglected crops, leafy vegetables, food security

1. Introduction

The population of the world is expected to reach 9.1 billion by 2050, and the larger percentage of this growth is expected to come from the low-income countries such as those in sub-Saharan Africa [1]. Consequently, this will result in more water competition in these countries, which are already frequently affected by drought stress and food insecurities. In South Africa, the rainfall is uneven throughout the country and often unpredictable and erratic. The country as a result has been characterized as the "water scarce country" with less than 500 mm of rainfall received per annum as compared with the world average of 836 mm [2]. The predicted effects of climate change in sub-Saharan Africa and the country will make the situation worse due to the predicted increase and intensity and frequency of drought stress [3]. Therefore, this is extremely worrying when looked within the context this will have in agricultural production and the vulnerability of the poor rural household and the poor urban, because the incident of crop failure will likely rise resulting in food insecurities within these communities [4]. This suggests that there is a need to continuously increase crop production and productivity by developing drought tolerant genotypes that are adapted to grow under varying and unpredictable conditions and under water scarcity conditions. This has resulted in the renewed efforts to consider the "landraces" as possible crops, which can safely guard food security in these marginal areas characterized by abiotic (drought and heat stress) and biotic (pest and diseases) environmental stress [5].

A review [6] on nutritional status of South Africans has revealed that under- (underweight, thinness, and stunting) and over-nutrition (obesity and overweight) and diet-related non-communicable diseases [7, 8] coexisted within the same communities and often on the same household. This phenomenon is known as the triple burden of malnutrition [9]. This has been exacerbated by the Covid-19 pandemic in 2020 and 2021 in South Africa where the government has provided R350 grants for the poorest affected people. Reportedly, the world hunger rose further in 2021 following a sharp upturn in 2020 during the Covid-19 pandemic making worse the existing inequalities, which have contributed to setbacks toward achievement of Zero Hunger Target by 2030 [10].

The increase in production and intake of traditional leafy vegetables was identified as the potential solution for addressing the problem associated with poor food security and nutritional imbalances in the rural communities [11, 12]. In contrast to urban and semi-urban communities, people in rural areas have access to land, which they can use to cultivate crops, thus contributing to their food security. However, this is usually hampered by a lack of adequate resources in these areas. Information, expertise, and low capital to buy agricultural inputs such as seeds, fertilizers, herbicides, and insecticides, as well as infrastructure needed to produce these crops, are often limiting in these areas. Under such circumstances, it has been suggested that traditional crop species can play a vital role in ensuring food security under such low input systems because of their likely suitability to these areas [13–15]. However, most traditional crops remain underutilized despite reports that they may be better suited to low input systems. This could be due to lack of clear policy instruments encouraging cultivation of these crops, lack of research interest from agricultural scientists, and low yields due to poor agronomic practices. Therefore, to promote the use of traditional crop species, there is a need to conduct research that will contribute to the documentation of optimum agronomic practices of these crops. This would contribute significantly to food security through increasing productivity of these crops, thus promoting balanced diets. Bottle gourd (*Lagenaria siceraria* (Molina) Standley) is one such crop species, which has been under researched in South Africa.

Bottle gourd is a member of the *Cucurbitaceae* family together with pumpkins and watermelons [16]. The crop originates in Africa, and it exhibits a great diversity in nature, and this alone indicates wide environmental adaptation [17]. The leaves of the crop are consumed the same way as those of pumpkins, watermelons, and other popular cucurbits. They are usually consumed as a relish with maize staple. The seeds of the crop, on the other hand, are popular snacks in Africa and are reported to contain high levels of proteins as is the case with the seeds of its closest relative pumpkins [18]. The mature fruit can be used as a container to store water, food, and as a musical instrument [19]. In addition, in Asia, bottle gourd is used as a rootstock in winter production of watermelons and squashes to prevent root-borne pathogens such as

Fusarium oxysporum [20]. A lot has also been documented about its medicinal properties, especially in countries such as India and Pakistan. Bottle gourd has been reported to contain high levels of choline, which is a compound that is reported to heal mental disorders [21]. In India, it has been reported to cure stomach complications [19].

Given all these benefits, it is important to note that the potential of bottle gourd landraces as a possible food security crop has been overlooked by many researchers. These landraces have been preserved by the communities who have been utilizing them for over 100 years and form a possible germplasm resource. They may have adapted to ecological niches from which they have been preserved. This makes them an important food security crop for cultivation in marginal areas of crop production. However, owing to the popularity of exotic members of the *Cucurbitaceae* family (pumpkins, watermelon, butternut, and squashes), the popularity and cultivation of bottle gourd landraces have faced neglect. Therefore, is a need to conduct research on improvement of this crop and their potential to contribute to food security in marginal areas of crop production. Therefore, the objective of this chapter is to review the research conducted on bottle gourd landraces and highlight their potential contribution to food security.

2. Water scarcity, food production, and climate change

The climate of the sub-Saharan region is characterized as semi-arid to arid with most countries receiving the rainfall that is less than 500 mm per annum. Frequent occurrence of drought is common in most countries in the region. This in combination with climate change and variability has led to huge losses in local food production and has threatened food security [22]. This is important to note since larger part of the population in sub-Saharan Africa is still dependent of agriculture for subsistence. This is a vulnerable group with low income or no income at all, low adaptive capacity to climate change, and facing a great treat to food security [23, 24]. The predicted impact of climate change in sub-Saharan Africa is that of rising temperature, increased incidence of floods, increased occurrence of extreme events such as drought, and increased rainfall variability [25]. Already, some of these conditions are experienced in sub-Saharan Africa with the recent case of severe floods that have affected many communities in Durban, South Africa, in 2021.

In sub-Saharan Africa, most economies are still dependent on agriculture as the driver of rural and economic development, and it is argued that current agricultural activities within these communities are too mainstream and lack the necessary innovation to allow rural economic development [24]. Reportedly, 95% of the population in the region is still dependent on rainfed agriculture, and it is subsistence-based [26]. This, therefore, is a huge challenge when viewed with the context of impact this will have on agricultural production and vulnerability of the smallholder farmers [22]. Therefore, given these challenges, there is a need to embrace new paradigms or alternative approaches that promote context-specific, best bet agricultural technology that can perform under water scarcity, climate change and provide food and nutritional security for the rural household. Those solutions should be practical, sustainable, and resilient to the problems facing sub-Saharan Africa such as water scarcity, climate change, and food security. As a result, landraces or underutilized crop species such as bottle gourd have been promoted in South Africa to cope with water scarcity, climate change, and food security.

3. Neglected underutilized indigenous crop species

In context of the above discussion, neglected and underutilized crop species have been recommended to mitigate some of these challenges. According to a report [27], neglected and underutilized crop species can be defined as those crops that have become indigenized over many years (more than 100 years) cultivation and natural and farmers selection within South Africa, thus they have become "landraces." On the other hand, landraces have been defined as the local variety of a plant species that evolved largely through selection by farmers in an unstructured way and which has become adapted to ecologies where it grows and survives [28]. These crops are sometimes called forgotten crops, neglected crops, orphans crops, underutilized crops, etc., as there is no consensus on what these crops should be referred to [22]. In the context of this document, landraces and neglected and underutilized crop species will be used interchangeable and have the same meaning. These crops currently have low level of use that is limited to subsistence farmers, have previously under-researched, and have not been previously classified and the major crops [29]. These crops previously, played an important role in ensuring household food security through providing alternative food source when the main crop has failed [29]. These crops also provided food source during times in-between subsequent harvest. Neglected crops species are cultivated from landraces using unimproved landrace varieties mostly by subsistence farmers. They are believed to be adapted to a range of ecological niches, both biotic, pest, and diseases and abiotic drought and heat stress. Therefore, there is a need to conduct more research on these crops and identify the desired traits as the initial stage of crop improvement. Therefore, these crops serve as an important germplasm for future crop improvements [22].

Promotion of few crops such as maize, wheat, soybeans in the mainstream agriculture has occurred at the expense of neglected indigenous crop species. Generally, these crops have been ignored in crop improvement programs by agronomist and plant breeders, and thus, they have earned the term "underutilized" [5]. There is currently little information describing their performance and responses to biotic and abiotic environmental conditions. The promotion of these crops will depend mainly to the availability of information about their growth and development, and bottle gourd is one such crop.

Bottle gourds exhibit large phenotypic variation, and this alone indicates wide genetic variation within the species. It is widely cultivated in sub-Saharan Africa for multiple uses by smallholder farmers [30, 31]. The crop is cultivated for its fruits, leaves, and seeds, and this can be consumed by both humans and livestock. The wide genetic variation within the crop provides the opportunity for selection and ideotype breeding [30]. There are no commercial varieties available of the crop in the region because of the lack of crop improvement programs dedicated into the crop. The crop is therefore classified as neglected underutilized crop species because its cultivation is mostly done by smallholder farmers using landraces under low input system with no improved seed varieties.

4. Drought tolerance of bottle gourd landraces

Larger percentage of the area in sub-Saharan Africa is characterized as arid to semi-arid. This therefore necessitate more studies on crop response to drought stress if the objective is to promote the neglected underutilized crop species. Drought or water stress in one of the single important environmental factors limiting crop production in the region. Drought can be defined as an extended period (months or years) where a region experiences a deficiency in water supply whether surface or underground water [32]. The plant mechanism to cope with stress is based on the choice of response it adopts in responding to developing water stress [22]. The responses of plants to water stress are complex [33]. The plant may tolerate, avoid, or escape drought stress [34], and the paragraph below sought to explain these mechanisms.

Drought avoidance is associated with reduction in water loss while enhancing or maintaining the uptake by the roots [22]. Drought escape on the other hand is associated with timing of key phenological stages. Crops can escape drought by having short growing season, and this allows them to complete their life cycle before water stress becomes terminal, and flowering time is an important trait related to drought escape [35]. While drought avoidance involves the responses of crop such as the regulation of stomata and the enhanced water capture, proliferation, and extensive root system [36]. Root thickness, root length, and root depth are some of the characteristics associated with enhanced water capture during drought stress [37]. Plant morphological changes such as number of leaves, plant height, leaf area, and leaf area index are reduced during drought stress, thereby assisting the plant to lose less water (or avoid drought) during water limiting conditions [38]. However, these mechanisms do not occur in isolation as drought avoidance can be associated with reduced leaf number and reduced season period, which is a trait of drought escape [34]. Finally, drought tolerance, which has been defined as the capacity of the plant to maintain metabolic activities under water scarcity or drought stress [34]. This is different from drought escape in that it does not show signs or evidence of yield reduction [34]. Drought tolerance is characterized by antioxidant defense system (i.e., reactive oxygen species and reactive nitrogen species) and osmotic adjustment (i.e., proline) [39]. Therefore, the understanding of all these mechanisms involved in water stress is important for strategic breeding of bottle gourd landraces and the release of improved varieties of the crop.

Bottle gourd in sub-Saharan Africa is grown by smallholder farmers from landraces. These landraces are believed to be adapted to stressful conditions of arid and semi-arid characterized by water stress and salinity. The recent studies conducted in South Africa on the response of the crop to water stress have shown some level of tolerance to these conditions [40–42] as compared to other cucurbits crops. As a result, it is suggested [31] that strategic crosses between drought-tolerant landrace accessions may enhance the level of drought tolerance and development of well-adapted hybrid varieties. This coupled with the understanding of the physiological mechanisms to drought tolerance can assist in effective screening and identification of the novel genes in the crop for breeding purposes. The accumulation of secondary metabolites such as cucubitacins during drought stress has been observed in the crop [40, 41] and can be associated with drought tolerance in bottle gourd landraces. In contrast, the concrete evidence that suggests their role in conferring drought tolerance is limited in cucurbit crops including bottle gourd landraces. Cucurbitacins E and I have been found to increase in response to water deficit [42], and the authors suggested that this may be the potential physiological marker of selection and identification of the crop for drought tolerance breeding. Thus, full understanding of the function of cucurbitacins during water deficit may assist in identification of the novel genes that confer drought tolerance [31].

5. Economic and medicinal uses

In South Africa, about 30 years ago, the crop was planted for the pipe making industry [43]. The necks of the fruits were bent in such a way that they grow to form bowls of pipes, and when the fruits were dry and ripe, the necks sawn off, cleaned, and exported to pipe markets overseas [43]. In Malaysia, rural farmers have been reported to increase their income due to planting and selling bottle gourd leaves and fruits [44]. A lot has been reported on medical properties of the crop. Bottle gourd has been shown to contain triterepenoide cucurbitanics B, D, G, and H, two sterols (fucosterol and campesterol), aerpene, byonolic acid, flavone-C glycosides, and lagenin [19]. The extract from the seed was found to contain antibiotic properties, and the fruit juice is helpful in constipation, premature graying hair, urinary disorder, and insomnia. To date, the crop has been found to contain high levels of choline [19, 21]. Choline serves as a precursor of the neurotransmitter acetylecholine, which in turn is important for retaining and enhancing memory. Further, a report [19] indicated that bottle gourd juice helped to regulate blood pressure in hypertensive patients because of its high potassium content, reducing weight quickly because of its high diet fiber and low fat and cholesterol content. Furthermore, the crop has been reported to lower blood cholesterol. In addition, the antihyper lipidemic effects (anti-lipids effect) of four different extracts have been explored [45]: chloroform, petroleum ether, alcoholic and aqueous extract from bottle gourd. They found that both chloroform and alcoholic extract had a significant effect on lowering total cholesterol, triglycerides, and low-density lipoprotein along with increase in high density lipoprotein as compared to others. Their results also suggested marked antihyperlipidemic and hypolipidemic effects of the extract.

6. Potential contribution to food security

In most parts of the world and in South Africa, bottle gourd is grown mainly as a vegetable for human consumption. Leaves of the crop are consumed the same way as those of popular cucurbits (watermelon, pumpkin, and squashes) and other popular leafy vegetables (spinach, Amaranthus spp., spider flower, chines cabbage, etc.). The young fruit of the crop is a popular vegetable in many parts of the world [46]. According to others [17], the leaves of the crop can also be added fresh and mixed with maize porridge in southern Africa, and they can also be dried and stored for later use in the off season. The seeds of the crop have been reported to contain high levels of oil that is comparable to those of sunflower and grape oil [47]. Other reports [48] indicated that bottle gourd was rich in protein, oil, and energy. Apart from these nutritional uses, bottle gourd has been used for decades in Asia as a root stock for watermelon to promote the root system under stressful conditions of water deficit and salinity [49], low temperature [50], as well as root-borne pathogens (Han et al., 2004 use numbers to cite references). In South Africa and the neighboring countries (Botswana and Zimbabwe), the oil is extracted from the seed and used as an alternative for vegetable oil [51]. According to other reports [17], the defatted cake can be used as a protein supplement in rural communities. *L. siceraria* seeds have been reported to contain about 45% oil and 35% proteins. [46] argued that the potential for bottle gourd as a food security crop lay on the use of its seed kernel in food and livestock industry. Given such benefits of the crop, it is a wonder that in South Africa, the benefits of the crop have not yet been fully exploited. The country is still faced with the problem of malnutrition. In recent years, obesity has increasingly become problematic in both rural and urban communities. High incident of stunted, underweight children, and increasing infant mortality due to marasmus and Kwashiorkor have been reported [17]. The use of bottle gourd seeds or defatted cake could help in mitigating protein deficiencies in rural communities. Oil extracted from bottle gourd is reported to be rich in sterolic compounds and fatty acids [47]. Thus, the use of bottle gourd seeds could contribute significantly in providing much needed amino acids in the diets of vulnerable communities. Nutritionists have argued that inclusion of leafy vegetables in diets could increase dietary diversity, nutrient availability, and absorption contributing to the reduction of malnutrition [52]. A report [19] indicated that the crop to form excellent diet that is rich in iron, vitamins, and minerals. In addition, the seeds and fruit of the crop can also be used to supplement livestock feeds in rural communities where grazing land is also becoming a problem. Observations in Zimbabwe have shown that leaves, seeds, and fruits are being used to supplement livestock feeds [53]. It has also been reported [17] that bottle gourd also contained sodium, potassium, essential elements, and trace minerals. They concluded that the crop could be useful to hypertensive patients since it contained high levels of potassium and sodium. Global population and that of South Africa continue to increase rapidly. This necessitates the production of more food to meet the growing demand by the increasing population. To achieve this, it is necessary for the country to look at the diversity that exists in traditional crops. Another problem that the country and the world face is that of climate change. It is important therefore to also look for diverse crops that can withstand high temperatures and possible outbreak of diseases in order to ensure food security, especially in the marginal areas of crop production. This makes bottle gourd attractive for a range of uses.

7. Conclusions

Bottle gourd landrace is an underutilized crop with a huge potential to contribute to food security. This crop has been growing in marginal areas of crop production characterized by water stress and salinity for hundreds of years, and thus, it is believed to be adapted to these conditions. The crop can strive well in the field with minimal agricultural inputs such as water and fertilizer and the use of herbicides and pesticides. The crop contains beneficial nutrients and medicinal properties that are beneficial to humans. The crop seeds have also been used in livestock supplements where the seed served as the protein source. Therefore, with the countries in sub-Saharan Africa faced with the problem of malnutrition, it is important to investigate the diversity of crops that are suited in these areas characterized by drought stress while at the same time providing enough nutrition to safeguard food security.

Conflict of interest

The authors declare no conflict of interest.

Author details

Nkanyiso J. Sithole[1*], Khayelihle Ncama[2], Lembe S. Magwaza[3,4*], Nozipho M. Motsa[5] and Jacob Mashilo[6]

1 Faculty of Natural and Agricultural Sciences, Department of Crop Sciences, North-West University, Mmabatho, South Africa

2 Faculty of Natural and Agricultural Sciences, Food Security and Safety Niche Area, North-West University, Mmabatho, South Africa

3 Discipline of Crop Science, School of Agricultural, Earth and Environmental Sciences, University of KwaZulu-Natal, Pietermaritzburg, South Africa

4 Horticultural Science, School of Agricultural, Earth and Environmental Sciences, University of KwaZulu-Natal, Pietermaritzburg, South Africa

5 Faculty of Science and Agriculture, Department of Agriculture, University of Zululand, Empangeni, South Africa

6 Limpopo Department of Agriculture and Rural Development, Towoomba Research Centre, Agriculture Regulatory and Technology Development, Crop Science Directorate, Bela-Bela, South Africa

*Address all correspondence to: nkanyisoj@gmail.com; justice.sithole@nwu.ac.za and magwazal@ukzn.ac.za

References

[1] United Nations. Department of Economics and Social Affairs, Population Division. World Population Prospects: The 2008 Revision. New York: United Nations Publications; 2009

[2] Department of Water Affairs and Forestry (DWAF). National Water Resource Strategy: Report. 1st ed. Pretoria, South Africa: Department of Water Affairs and Forestry (DWAF); 2002

[3] Hassan R. Climate change and African agriculture. Policy No. 28. Based on Durand (2006), "assessing the impact of climate change on crop water use in South Africa". In: CEEPA Discussion Paper No. 28, CEEPA. South Africa: University of Pretoria; 2006

[4] Sisulu S, Scaramella C. Food for Thought: A Climate for Ending Hunger. Skyways, February 2012. Kyalami, South Africa: Panorama publications; 2012. pp. 31-33

[5] Mabhaudhi T. Drought Tolerance and Water-Use of Selected South African Landraces of Taro (*Colocasia esculenta* L. Schott) and Bambara Groundnut (*Vigna subterranea* L. Verdc). PhD thesis. Pietermaritzburg, South Africa: Crop Science, University of KwaZulu-Natal; 2012

[6] Oelofse A, van Averbeke W. Nutritional value and water use of African leafy vegetables for improved livelihoods. Water Research Commission Report No TT 535/12. 2012:294-302

[7] Govender L, Pillay K, Siwela M, Modi A, Mabhaudhi T. Food and nutrition insecurity in selected rural communities of KwaZulu-Natal, South Africa-linking human nutrition and agriculture. International Journal of Environmental Research and Public Health. 2016;**2016**(14):17. DOI: 10.3390/ijerph14010017

[8] Dakora FD, Belane AK. Evaluation of protein and micronutrient levels in edible cowpea (*Vigna Unguiculata L.* Walp.) leaves and seeds. Frontiers in Sustainable Food System. 2019;**3**:70. DOI: 10.3389/fsufs.2019.00070

[9] Modjadji P, Madiba S. The double burden of malnutrition in a rural health and demographic surveillance system site in South Africa: A study of primary schoolchildren and their mothers. BMC Public Health. 2019;**19**:1087. DOI: 10.1186/s12889-019-7412-y

[10] FAO, IFAD, UNICEF, WFP, WHO. The state of food security and nutrition in the world 2022. In: Repurposing Food and Agricultural Policies to Make Healthy Diets more Affordable. Rome: FAO; 2002. DOI: 10.4060/cc0639en

[11] Schippers RR. African Indigenous Vegetables. An Overview of the Cultivated Species 2002—Revised Version on CD-ROM. Aylesford, UK: Natural Resources International Limited; 2002

[12] Laker MC. Introduction to the special edition of water SA on indigenous crops, water and human nutrition. Water SA. 2002;**33**:317-326

[13] Modi M, Modi AT, Hendriks S. Potential role of wild vegetables in household food security: A preliminary case study in KwaZulu-Natal, South Africa. African Journal of Food Agriculture Nutrition and Development. 2006;**6**:1-13

[14] Schonfeldt HC, Pretorius B. The nutrient content of five traditional south

African dark vegetables-a preliminary study. Journal of Food Composition and Analysis. 2011;**24**:1141-1146

[15] Odhav B, Beekrum S, Akula U, Baijnath H. Preliminary assessment of nutritional value of traditional leafy vegetables in KwaZulu-Natal, South Africa. Journal of Food Composition and Analysis. 2007;**20**:430-435

[16] Decker-Walter DS, Wilkins-Ellert M, Chung S, Staub JE. Discovery and genetic assessment of wild bottle gourd [*Lagenaria siceraria* (Mol.) Standley; Cucurbitaceae] from Zimbabwe. Economic Botany. 2004;**58**:501-508

[17] Chimonyo VGP, Modi AT. Seed performance of selected bottle gourd (*Lagenaria siceraria* (Molina) Standl.). American journal of. Experimental Agriculture. 2013;**3**:740-766

[18] van Wyk BE, Gericke N. People's Plants: A Guide to Useful Plants of Southern Africa. Pretoria, South Africa: Briza Publications; 2001. p. 325

[19] Milind P, Satbir K. Is bottle gourd a natural guard? International Research Journal of Pharmacy. 2011;**6**:13-17

[20] Han JS, Mok IG, Park HG, Kim CK. Efficient plant regeneration from the cotyledon explants of bottle gourd (Lagenaria siceraria Standl.). Plant Cell Reports. 2004;**23**:291-296. DOI: 10.1007/s00299-004-0846-3

[21] Rahman ASH. Bottle gourd (Lagenaria siceraria): A vegetable for good health. Natural Product Radiance. 2003;**2**:249-254

[22] Chivenge P, Mabhaudhi T, Modi AT, Mafongoya P. The potential role of neglected and underutilised crop species as future crops under water scarce conditions in sub-Saharan Africa. International Journal of Environmental Research and Public Health. 2015;**12**:5685-5711. DOI: 10.3390/ijerph120605685

[23] Jones L, Carabine E, Roux JP, Tanner T. Promoting the Use of Climate Information to Achieve Long-Term Development Objectives in Sub-Saharan Africa: Results from the Future Climate for Africa Scoping Phase. Oxford, UK: Future Climate For Africa; 2015

[24] Mudau FN, Chimonyo VGP, Modi AT, Mabhaudhi T. Neglected and underutilised crops: A systematic review of their potential as food and herbal medicinal crops in South Africa. Frontiers in Pharmacology. 2022;**12**:809866. DOI: 10.3389/fphar.2021.809866

[25] International Panel on Climate Change (IPCC). Working Group 1 Fourth Assessment Report "The Physical Science Basis". 2017. Available online: http://www.ipcc-wg1.unibe.ch/publications/wg1-ar4/wg1-ar4.html. [Accessed: May 22, 2017]

[26] Singh P, Wani SP, Pathak P, Sahrawat KL, Singh AK. Increasing crop productivity and water use efficiency in rainfed agriculture. In: Integrated Watershed Management; Wani, SP, Rockstrom J, SaHrawat L, Eds.; London, UK: CRC Press; 2011; charpter 10, p. 315-348

[27] Schippers RR. Traditional vegetables in Africa. In: Proc. Int. Symp. On the Nutritional Value and Water Use of Indigenous Crops for Improved Livelihoods. (Available on CDROM). Pretoria, South Africa: Univ. Pretoria; 2006

[28] Nass LL, Parterniani E. Pre-breeding: A link between genetic resources and

maize breeding. Scientia Agricola. 2000;**57**:581-587

[29] Azam-Ali SN. Fitting underutilized crops within research-poor environments: Lessons and approaches. South African Journal of Plant and Soil. 2010;**27**:293-298

[30] Mashilo J, Shimelis H, Odindo A. Genetic diversity of bottle gourd (*Lagenaria siceraria* (Molina) Standl.) landraces of South Africa assessed by morphological traits and simple sequence repeat markers. South African Journal of Plant and Soil. 2015;**33**:113-124. DOI: 10.1080/02571862.2015.1090024

[31] Mkhize P, Mashilo J, Shimelis H. Progress on genetic improvement and analysis of bottle gourd [*Lagenaria siceraria* (Molina) Standl.] for agronomic traits, nutrient compositions, and stress tolerance: A review. Frontiers in Sustainable Food Systems. 2021;**5**:683635. DOI: 10.3389/fsufs.2021.683635

[32] Oval Myers J, John HY, Krishnamani MRS. Breeding soyabean for drought resistance. Plant Breeding. 1986;**4**:203-232

[33] Blum A. Drought resistance – It is really a complex trait? Functional Plant Biology. 2011;**38**:753-757

[34] Blum A. Drought resistance, water-use efficiency, and yield potential – Are they compatible, dissonant, or mutually exclusive? Australian Journal of Agricultural Research. 2005;**56**:1159-1168

[35] Araus J, Slafer GA, Reynolds MP, Royo C. Plant breeding and drought in C3 cereals: What should we breed for? Annals of Botany. 2002;**89**:925-940

[36] Kavar T, Maras M, Kidric M, Sustar-Vozlic J, Meglic V. Identification of genes involved in the response of leaves of Phaseolus vulgaris to drought stress. Molecular Breeding. 2007;**21**:81-87

[37] Turner NC, Wright GC, Siddique KHM. Adaptation of grain legumes (pulses) to water-limited environments. Advances in Agronomy. 2001;**71**:123-231

[38] Mitchell J, Siamhan D, Wamala MH, Risimeri JB, Chinyamakobvu E, Henderson SA, et al. The use of seedling leaf death score for evaluation of drought resistance of rice. Field Crops Research. 1998;**55**:129-139

[39] Farooq M, Wahid A, Kobayashi N, Fujita D, Basra SMA. Plant drought stress: Effects, mechanisms and management. Agronomy for Sustainable Development. 2009;**29**:185-212

[40] Sithole NJ, Modi AT. Responses of selected bottle gourd [*Lagenaria siceraria* (Molina Standly)] landraces to water stress. Acta Agriculturae Scandinavica, Section B — Soil and Plant Science. 2015;**65**:350-356

[41] Mashilo J, Odindo AO, Shimelis HA, Tesfay SZ MP, Magwaza LS. Drought tolerance of selected bottle gourd [*Lagenaria siceraria* (Molina) Standl.] landraces assessed by leaf gas exchange and photosynthetic efficiency. Plant Physiology and Biochemistry. 2017;**120**:75-85

[42] Mashilo J, Odindo AO, Shimelis HA, Musenge P, Tesfay SZ, Magwaza LS. Photosynthetic response of bottle gourd [*Lagenaria siceraria* (Molina) Standl.] to drought stress: Relationship between cucurbitacins accumulation and drought tolerance. Scientia Horticulture. 2018;**231**:133-143

[43] Welman M. National Herbarium, Pretoria. South African National Biodiversity Institute (SANBI); 2005.

Also available on line: http://www.plantzafrica.com/plantklm/lagensic.htm. Accessed July 30, 2022

[44] Awasthi BD. Relevance of Market Information System to Environment Protection. Ministry of Agriculture and Co-operatives: Singhdurbar; 2012. pp. 46-54

[45] Ghule VB, Ghate MH, Saoji AM, Yeole PG. Hypolipidemic and antihyperlipidemic effect of Lagenaria siceraria (Mol.) fruit extracts. Indian Journal of Experimental Biology. 2006;**44**:905-909

[46] Prasad R, Prasad A. A note on the heritability and genetic advance in bottle gourd (*Lagenaria siceraria* (Mol.) Standl.). Indian Journal of Horticulture. 1979;**36**:446-448

[47] Axtell BL, Fairman RM. Minor Oil Crops: Edible oils Food and Agriculture Organization of the United Nations. 1992. Available: http://www.fao.org/docrep/X5043E/X5043E00.htm. [Accessed: September 10, 2022]

[48] Loukou AL, Lognay G, Barthelemy JP, Maesen P, Baudoin JP, Zoro BI. Effect of harvest time on seed oil and protein contents and compositions in the oleaginous gourd *Lagenaria siceraria* (Molina) Standl. Journal of Science Food and Agriculture. 2011;**91**:2073-2080

[49] Park M, Won H, Kim J, Han J, Ahn Y. Expression of H+ phosphatase from Arabidopis enhances resistance of drought stress in transgenic bottle gourd (*Lagenaria siceraria* Standl.). In: Sari, Somlaz, Aras, editors. Proceedings of Xth EUCARPIA Meeting on Genetics and Breeding of Cucurbitaceae. Antalya (Turkey): University of Cukurova, Ziraat Fakultesi; 2012

[50] Yetisir H, Sakar M, Serce S. Collection and morphological characterization of Lagenaria siceraria germplasm from the Mediterranean region of Turkey. Genetic Resource Crop Evolution. 2008;**55**:1257-1266

[51] Grubben GJH, Denton OA. Plant Resources of Tropical Africa 2. Vegetables. Wageningen, Netherlands: PROTA foundation, Wageningen, Netherlands/Backhuys Publishers, Leiden, Netherlands/CTA; 2004

[52] Maunder EMW, Meaker JL. The current potential contribution of home-grown vegetables to diets in South Africa. Water SA (Special Edition). 2007;**33**:401-406

[53] Chikwanha OC. Seasonal Changes in Body Condition Scores of Pigs and Chemical Composition of Pig Feed Resources in a Semi-Arid Smallholder Farming Area of Zimbabwe. MSc thesis,. Harare, Zimbabwe: Department of Animal Science, University of Zimbabwe; 2006

Chapter 6

Study of Genetic Resources of Winter Wheat and Identification of Genetic Sources with Drought Resistance for Use in Breeding and Production

Sulukhan Temirbekova, Ivan Kulikov, Natalya Ionova, Yuliya Afanasyeva and Elena Kalashnikova

Abstract

The problem of drought is acute in a large area of Russia, which will not decrease in the coming decades but will grow. The most important measures to combat drought are the selection of drought- and heat-resistant crops and the creation of varieties for various ecological and geographical zones of Russia based on the widespread use of the World Collections of Agricultural Plants. The results of the study of the gene pool of winter wheat during severe atmospheric drought, which manifested itself twice in 50 years of studying genetic resources from different countries in 1972 and 2010, are presented. The primary tasks were the mobilization of new forms of drought-resistant plants from the arid zones of Russia, as well as from abroad, the expansion of research on the identification of genetic sources and donors of drought resistance, the creation and accelerated introduction into agricultural production of new drought-resistant varieties and hybrids of winter wheat.

Keywords: winter wheat, genetic sources, drought resistance, breeding, production

1. Introduction

The drought problem is serious in most areas of our country. In addition, as the results of research by the largest Russian and foreign climate scientists show [1], the probability of this adverse phenomenon will not decrease in the coming decades but will increase. N.I. Vavilov [2] rightly considered the selection of drought- and heat-resistant crops and the creation of drought-resistant cultivars for various ecological and geographical zones of the country based on the extensive use of the All-Union Institute of Plant Growing agricultural plants world collections will be the most important measures to cope with drought.

For the first time, the classification of agricultural plants cultivated in the USSR according to the degree of drought resistance was made by N.I. Vavilov and reported at the All-Union Conference on problems related to drought, held by the USSR Academy of Sciences and VASHNIL in 1931 in Moscow [2]. The need for such work, N.I. Vavilov notes in his report, that is related to the special agricultural conditions in our country, characterized by the widespread and frequent occurrence of such an unfavorable phenomenon as drought.

All the varieties of plant species and genera by agro- and ecological specifications were divided into three groups by N.I. Vavilov [2].

The first group included plants that are more resistant to drought and capable of yielding crops even in the conditions of an acutely arid summer. These, along with such xerophytic plants, such as cactus and agaves, include millet, sorghum, chickpeas, fine-grained lentils, various types of lathyrus, sweet clover, granary, mogar, yellow alfalfa, sudan grass, sheep fescue, and a number of fruit and essential oil crops.

The second group consists of plants with intermediate resistance, having a large range of variability and showing a relative resistance to drought, able to produce crops with minimum moisture. Plants of this group are of the greatest importance in agriculture, occupying a significant part (more than 3/5) of the agricultural area of crops. It includes crops such as wheat, barley, maize, rye, sunflower, cotton, sugar beet, alfalfa, and vicia.

The third group includes plants less resistant to drought, which can yield only in conditions of sufficient moisture. This group consists of the vast majority of plants grown in the territory of our country.

The study of the world's variety of agricultural plants has shown that the most precious source material resistant to drought is collected in the territory of our country. The priority tasks at the present stage are mobilization of new forms of drought-resistant plants from the arid zones of our country, as well as from abroad; expanding research into identifying genetic sources and donors of drought resistance; accelerated introduction of new drought-resistant varieties, and the most important crop hybrids into agricultural production.

2. World gene pool of winter wheat for drought resistance

It was a severe atmospheric drought in 1972 and 2010, during 50 years studying of the winter wheat gene pool from the world collection of VIR in the conditions of the Moscow region.

In 1972, grain maturation took place under conditions of exceptional atmospheric drought. The summer of 1972 lasted 145 days, from May 2 to September 24, and was characterized by the predominance of hot and sunny weather, high-temperature stress, and high water deficit.

The average monthly air temperature in May was 11.8°C, which is not significantly higher than normal, but this period was characterized by a lower than average level of monthly precipitation: 28.6 mm precipitation rate fell although the average monthly rate—50 mm. In combination with the practical absence of atmospheric precipitations in the second and third decades of April, this lead to a significant deficit, which negatively affected the synchrony and speed of spring cereal seedlings, their growth, and development, in leguminous crops and perennial grasses.

The water content in the first one-meter layer of soil under wheat decreased by 26 mm in May—from 209 to 183. For 7 days, the relative humidity was below 40%.

The average monthly temperature in June was 18.7°C, which is 3.3°C lower than the average long-term norm. The absolute maximum has reached 32.0°C and the minimum has not decreased below 9°C.

The amount of atmospheric precipitations in June was only 16.4 mm with a standard of 72 mm, and due to their dispersal, the biological effectiveness of precipitation was very low. As a consequence, the reserves of productive water in the soil continued to decrease and amounted to 155 mm in the first one-meter layer of soil under wheat in the first decade of June, 135 in the second, and 126—in the third.

In July, further warming was observed; the average daily air temperature for the month was 22°C with a norm of 17.7°C. The absolute maximum in the first decade reached 35.5°C, in the second—34.0°C, in the third—32°C, and the absolute minimum has not dropped below 7°C. On the surface of the soil, the temperature reached 4550°C.

Against such a high-temperature background, the amount of precipitation was extremely small, only 14 mm of precipitation fell against the average monthly rate—99 mm. The increasing lack of moisture created a high soil drought. The moisture reserve in the first one-meter soil layer below the winter wheat was 90 mm in the first decade and 107 mm in the second.

Three days in a month, there was air dryness (relative humidity of no more than 30%) and 10 days with humidity in a range of 31–40%.

August weather conditions differed little from July. The average monthly air temperature was 20.3°C, which is 4.4°C higher than normal. The absolute maximum temperature reached 34.5°C in the first decade, 35.5°C in the second, and 36.0°C in the third, which exceeds this indicator for the last 92 years. The absolute minimum air temperature was 5.5°C. The amount of rain during the month was 18.1 mm with a standard of 76 mm.

The increasing lack of moisture contributed to the intensification of atmospheric drought, in August, 12 days with a relative humidity of 16–29% and 8 days with a humidity of 34–39% were recorded.

In such conditions, there were difficulties in sowing winter crops.

By the end of August, the sum of positive air temperatures since the beginning of the growing season was 2381°C with an average annual norm of 1932°C, and the amount of atmospheric precipitations rate was 177 and 298 mm, respectively.

It was extremely difficult to perform experimental and agricultural work under such weather conditions.

The air temperature in September was equal to the long-term average—10.3°C. The first two decades were characterized by drought, and in the third 52.4 mm of precipitation fell. They have significantly improved the conditions for the emergence of winter seedlings, very weak and sparse within a month from the start of sowing. Collection of nurseries of wheat and rye were flown in the second half of September. The moisture reserve in the first 1 m layer of soil under wheat was 145, 143, and 178 mm, respectively, for the decades of September.

During the assessing winter wheat from the VIR World Collection for drought resistance, we used mainly agronomic indicators: the yield of varieties and its decrease in drought conditions compared to the control. The mass of 1000 grains was also identified as one of the main agronomic indicators. In total, 500 samples from different countries of origin were studied.

The mass of 1000 grains. In the reporting year, grain loading took place in conditions of exceptional soil and atmospheric drought. Only thanks to the double loosening of the row spacing in the soil, an insignificant amount of moisture available to plants has been preserved. Wheat leaves had almost completely dried up by June 26, and further accumulation of dry matter occurred only through the root system and the stem, which was still green. As a result, the grain maturation continued, but its value turned out to be 30–40% less than in normal weather conditions years. The mass of 1000 grains of the standard cultivar—Mironovskaya 808 in 1972 was 36.1 g. It was noted that 49 samples (out of 300) in 1972 were approaching the norm on this indicator and 12 exceeded it (**Table 1**).

Productivity. In the conditions of 1972, a natural assessment of winter wheat samples for drought resistance was made. The standard cultivar yield was—240 g/m^2 (compared to 372 g in 1969 and 358 g in 1970).

Most of the samples from Germany had a yield lower than that of the zoned variety. A yield equal to the standard cultivar level had only one sample in 1972—Steiners strusi k-44858—246 g/m^2; a yield of 4 samples: Golland k-39583—228 g/m^2, Trankensteiner Brauner k-40914—220 g/m^2, Goldene Aue k-40477—216 g/m^2, and Stiegler 22 k-26353—216 g/m^2 was almost at the level of the standard cultivar—Mironovskaya 808—240 g/m^2. Other samples: Heines Teverson K-185, Shiriffs K-1672, Kujavischer weisser Kolben K-6290, St 3876 50 k-43054, Lolimanns Beseler III k-26403, Cimbals Grossherzog V Sachz k-26205, Bensings Trotzkopf k-26228, Continental Dickkopf K-26310, and Liessau K-26354 had yield from 204 to 214 g/m^2.

Under very dry conditions, in 1972, the following samples were identified by drought resistance and a set of signs:

1. Liessau (κ-26354)—is winter-hardy, quite productive, matures 1 day later than the standard Mironovskaya 808.

2. Heine Stamm 3256 (κ-40864)—is winter-hardy, matures at the same time as the standard, relatively short, resistant to lodging, has a fairly large grain, yielding.

3. Loosdorfer Prasident Hanisch (κ-40894)—matures at the same time as the standard, is stunted, and weakly affected by powdery mildew.

4. Heines 1751 (κ-41245)—is winter-hardy, matures at the same time as the standard, relatively stunted, resistant to lodging, and yielding.

VIR catalog number	Cultivar	Massof 1000 grains (g)
40469	Heinnriehsvon Heindenburg	40
40476	Konkurrenzenvon Meyer Wageninger	38
26208	Hildebrandts Weissweizen	39
40467	Heinrichs Gelbkoerniger Dickkopf	39
45029	Dippes Triumph	38
43034	Fanal	37
40468	Hildebrandts WeisserViktoria	37
39737	Bielers Edelepp	38
44973	Skumstall	37
40487	Halletspedigree v. vilmorin	37
44796	Basta	39
43920	Mironovskaya 808, st.	36

Table 1.
The most cultivated winter wheat cultivars in Germany (Germany, 1972) stand out for their drought resistance.

5. Steiners Strusi (к-44858)—matures at the same time as the standard, relatively short, weakly affected by powdery mildew, and quite yielding in arid conditions.

6. Bielers Edelepp (к-39737)—is winter-hardy, has a large grain, is quite yielding, and is weakly affected by powdery mildew.

7. 38/120 (к-40105)—stunted, resistant to lodging, and weakly affected by brown rust.

8. Neuzucht 14/14 (к-40109)—stunted, resistant to lodging, weakly affected by brown rust and powdery mildew, quite winter-hardy, and yielding.

9. Schindlers N. Z. (к-40472)—has a large grain and good yielding.

10. Russe 991 (к-40858)—matures at the same time as the standard, is stunted, weakly affected by brown rust.

11. Standerers Markus (к-35660)—is winter-hardy, has a large grain, and matures later than the standard for 1 day.

12. к-39751—quite winter-hardy, stunted, weakly affected by brown rust, and has a large grain.

13. Halle 1020 (к-34063)—weakly affected by powdery mildew, has a large grain.

Thus, the source material of winter wheat from Germany is of particular interest for practical use (**Table 2**).

Drought-resistant samples selected from varieties of other countries in 1972: local cultivar (k–25029) Uzbekistan—late ripening, resistant to lodging, plant height 115 cm, affected by brown and yellow rust to an average degree, mass of 1000 grains —32 g, grain yield from 1 m^2 to 130 g, while the yield of standard cultivar—240 g/m^2;

The local variety (k-36323) Turkmenistan is precocious, the plant height is 120 cm, resistant to lodging, the defeat of powdery and yellow rust is average, the mass of 1000 grains is 28 g, the grain yield from 1 m^2 is 170 g, while the yield of standard cultivar is 240 g/m^2;

Impeto (k-40296) Italy—precocious, plant height 90 cm, powdery and yellow rust is affected to an average degree, weight of 1000 grains—32 g, grain yield from 1 m^2 to 175 g, while the yield of standard cultivar is 240 g/m^2;

Tepas (k-44546) Italy—plant height 80 cm, resistant to lodging and brown and yellow rust, mass of 1000 grains—30 g, grain yield from 1 m^2 to 190 g, while the yield of standard cultivar is 240 g/m^2;

Local cultivar (k-12757) Afghanistan—precocious, plant height 110 cm, resistant to rust diseases, mass of 1000 grains—26 g, grain yield from 1 m^2 to 150 g, while the yield of standard cultivar is 240 g/m^2;

Local cultivar (k-24084) India—precocious, plant height 105 cm, resistant to lodging, good resistance to rust diseases, mass of 1000 grains—20 g, grain yield from 1 m^2 to 120 g, while the yield of standard cultivar is 240 g/m^2;

Monon (k-44397) USA—precocious, plant height 100 cm, resistant to lodging, mass of 1000 grains—30 g, grain yield from 1 m^2 to 195 g, while the yield of standard cultivar is 240 g/m^2;

VIR catalog number	Cultivar	Grain mass from 2 m^2, g
43920	Mironovskaya 808, st.	480
44858	Steiners Strusi	492
39583	Golland	456
40914	Frankensteiner Brauner	440
40477	Goldene Aue	432
26353	Stiegler 22	432
185	Heines Feverson	428
1672	Shiriffs	428
6290	Kujavischer Weisser Kolben	420
43054	St 3876/50	420
26403	Lohmanns Beseler III	412
26205	Cimbals Grossherzog V.Sachz	412
26228	Bensings Trotzkopf	408
26310	Berkners Continental Dickkopf 95	408
26354	Liessau	408

Table 2.
The most productive winter wheat cultivars in Germany in 1972.

Arache (k-45079) USA—precocious, height of plant—95 cm, resistant to lodging and brown and yellow rust, mass of 1000 grains—32 g, grain yield from 1 m^2 to 210 g, the yield of standard cultivar is 240 g/m^2;

Albidum 11 (k-46730) Russia—medium-ripened, created with the participation of wheat from California, plant height 90 cm, resistant to diseases, resistant to lodging, mass of 1000 grains—39 g, grain yield from 1 m^2 to 386 g, the yield of standard cultivar—240 g/m^2;

Bezostaya 1 (k-42790) Russia—precocious, resistant to lodging, height of plants—105 cm, affected by rust diseases to an average degree, mass of 1000 grains—37.8 g, grain yield from 1 m^2 to 200 g, the yield of standard cultivar—240 g/m^2;

Odesskaya 3 (k-38441) Ukraine—is medium-ripe, height of plants—95 cm, resistant to lodging, is affected by rust diseases to an average degree, the mass of 1000 grains is 35.1 g, the grain yield from 1 m^2 is 220 g, the yield of standard cultivar—240 g/m^2;

As a result, a comparative analysis of the parameters of drought resistance in the VIR World Collection of samples revealed genotypes with high resistance to drought from Germany (in the first place), single samples from Italy, the USA, and Russia. The selected samples are of value and interest for practical use in breeding. Breeders have created drought-tolerant and highly productive cultivars on their basis in the Russian Federation.

The second severe drought in the Russian Federation over the past 50 years occurred in 2010.

The vegetation conditions in 2010 were quite unfavorable. The average temperature of the vegetation season was 6.5°C (22.9°C) above the average long-term norm (16.4°C). In May 2010, weather conditions were favorable for the growth and development of winter wheat plants—the air temperature corresponded to the average annual value (14.2–14.5°C). The last rain fell on June 18 and there was no rainfall until September 3. The air temperature in the Moscow region in June was 33°C, in July—up to 38°C, in August—up to 39.7°C (the average long-term temperature was +18.2°C, +20.5°C, +19.0°C in June, July, and August, respectively), the hydrothermal humidification coefficient (HHC) was 0.8. The lack of precipitation, as well as the abnormally high temperature of the air, has created a threat to the normal development and maturation of the wheat grain. The full maturation of the winter wheat samples took place between July 10 and 15, almost a month earlier than the optimal period. In conditions of severe atmospheric drought, the collection was evaluated for drought resistance. A total of 500 samples from different countries were analyzed. Among these, 42 samples of winter wheat were characterized by drought resistance, including 14 high-yield samples (**Table 3**).

It should be noted that drought is one of the most complex and destructive global abiotic stressors associated with agriculture in many countries of the world. The resulting damage is greater than the damage caused by all other stressors. In the European part of Russia, over 50 years of study of the world genetic heritage of winter wheat, the drought happened twice: in 1972 and 2010. According to the time of onset and duration, the drought began from the middle to the end of the growing season. Some researchers have noticed the onset of drought at the beginning, middle, or end of the growing season, which had different levels of intensity [3–5]. Photosynthesis is impossible without water and its net productivity is strictly limited by the availability of available moisture in the soil. It should be noted that all cultivars, including drought-resistant cultivars, react negatively to water deficiency, which limits productivity.

Selection for drought resistance cannot be considered separately from crop production technology, the main task of which should be the accumulation and

VIR catalog number	Origin	Cultivar	Grain mass per 1 m^2, g
Standard	Moscow region	Moscovskaya 39, st.	475
64065	Germany	Taras	670
64062	Germany	Tarmer	665
57008	Germany	TAW 7032/74	655
54633	Moscow region.	Ferrugineum 737/76	640
64061	Germany	Taroz	680
54131	Sweden	SV 71536	630
55971	Kursk region .	L-1749	620
57222	Germany	Severin	620
54635	Moscow region.	Lutescens 181/75	615
54689	Moscow region.	Lutescens 12424/74	615
64054	Poland	Juma	615
55801	Kursk region .	Lutescens 12	600
55315	Sweden	WW 71919	600
55246	Sweden	Sture	600

Table 3.
High yield, drought-resistant winter wheat samples from VIR World Wheat Collection, 2010.

preservation of water in the soil. Since under long-term drought conditions, the yield has been determined and restricted by the pre-drying water resources in the soil.

Drought and high temperatures affect not only the yield but also the milling qualities of grain, as well as the baking properties of flour. At the same time, the size, exhaustiveness, weight of the grain, and the mass of flour also decrease [6, 7]. During these years, the protein content of grains and flour increases and reaches 17–19% or more, gluten—up to 45–50%. An extreme air temperature of over 34–35°C during the grain maturation period can alter the gene expression of various groups of proteins between glutenins and gliadins [7–10].

Global warming and the disastrous 2010 drought are worrying about the production of cereals and other agricultural products. Modern conditions demand cultivars that use water more efficiently not only in dry years but in wet years as well. The solution to this problem is impossible without the enrichment of wheat plants by the genes of ancient local cultivars, which control resistance to a complex of pests and plant pathogens, extreme temperature tolerance, high yield and grain quality, and crop production.

Traditional recombination has now been partially replaced by molecular selection methods, but it should not be forgotten that recombinant breeding remains the basis for the use of indirect molecular technologies for the creation of new cultivars and hybrids using the genetic heritage of the best genotypes from the VIR World Wheat Collection.

Weather conditions in 2010 demonstrated the suitability of winter wheat and spring wheat breeding for drought resistance.

Below we present, particularly valuable collection samples of the worldwide VIR collection for use in the selection process in different countries [11], distinguished by drought resistance. In Russia, they are already in the breeding process in different regions.

- к-64061, Taroz, (Germany); variety: *lutescens*

    Basic signs:

    1. Winter hardiness—9 points (high).
    2. Characteristics of the plants by development phases (points):
    3. before leaving for winter—9 points;
        a. after wintering—9 points;
        b. before harvesting—9 points.
        c. The height of the plants before harvesting—95 cm.
    4. Resistance to lodging—9 points.
    5. The mass of 1000 grains—43.6 g.
    6. The defeat of powdery mildew and brown rust—high.

7. Grain mass from 1 m^2 to 680 g/m^2.

8. Grain mass from 1 m^2 in comparison with the standard—141.8%.

9. Grain yield from1 hectare—68 centners per hectare (c/ha).

- к-64065, Taros, Germany, variety – *lutescens*.

  Basic signs:

1. Winter hardiness—9 points (high).

2. Characteristics of the plants by development phases (points):

   a. before leaving for winter—9 points;

   b. after wintering—9 points;

   c. before harvesting—9 points.

3. The height of the plants before harvesting—95 cm.

4. Resistance to lodging—9 points.

5. The mass of 1000 grains—43.5 g.

6. The defeat of powdery mildew and brown rust—high.

7. Grain mass from 1 m^2- 670 g/m^2.

8. Grain mass from 1 m^2 in comparison with the standard—139.6%.

9. Grain yield from 1 hectare—67 c/ha.

- к-64062, Tarmer, Germany, variety – *lutescens*.

  Basic signs:

1. Winter hardiness—9 points (high).

2. Characteristics of the plants by development phases (points):

   a. before leaving for winter—9 points;

   b. after wintering—9 points;

   c. before harvesting—9 points.

3. The height of the plants before harvesting—90 cm.

4. Resistance to lodging—9 points.

5. The mass of 1000 grains—42.4 g.
6. The defeat of powdery mildew and brown rust—high.
7. Grain mass from 1 m^2- 665 g/m^2.
8. Grain mass from 1 m^2 in comparison with the standard—138.5%.
9. Grain yield from 1 hectare to 66.5 c/ha.

- к-64060, Tazit, Germany, variety – *lutescens*.

  Basic signs:

1. Winter hardiness—9 points (high).
2. Characteristics of the plants by development phases (points):
    a. before leaving for winter—9 points;
    b. after wintering—9 points;
    c. before harvesting—9 points.
3. The height of the plants before harvesting—85 cm.
4. Resistance to lodging—9 points.
5. The mass of 1000 grains—42.7 g.
6. The defeat of powdery mildew and brown rust—high.
7. Grain mass from 1 m^2- 660 g/m^2.
8. Grain mass from 1 m^2 in comparison with the standard—137.5%.
9. Grain yield from 1 hectare -66 c/ha.

- к-57008, TAW 7032/74, Germany, variety – *lutescens*.

  Basic signs:

1. Winter hardiness—9 points (high).
2. Characteristics of the plants by development phases (points):
    a. before leaving for winter—9 points;
    b. after wintering—9 points;
    c. before harvesting—9 points.

3. The height of the plants before harvesting—95 cm.
4. Resistance to lodging—9 points.
5. The mass of 1000 grains—36.0 g.
6. The defeat of powdery mildew and brown rust—high.
7. Grain mass from 1 m^2 to 665 g/m^2.
8. Grain mass from 1 m^2 in comparison with the standard—136.3%.
9. Grain yield from 1 hectare to 66.5 c/ha.

- к-57581, Gama, Poland, variety – *lutescens*.

  Basic signs:

  1. Winter hardiness—9 points (high).
  2. Characteristics of the plants by development phases (points):
     a. before leaving for winter—9 points;
     b. after wintering—9 points;
     c. before harvesting—9 points.
  3. The height of the plants before harvesting—85 cm.
  4. Resistance to lodging—9 points.
  5. The mass of 1000 grains—40.7 g.
  6. The defeat of powdery mildew and brown rust—high.
  7. Grain mass from 1 m^2 to 650 g/m^2.
  8. Grain mass from 1 m^2 in comparison with the standard—135.4%.
  9. Grain yield from 1 hectare to 65 c/ha.

- к-54633, Ferrugineum 737/76, Moscow region, variety –*ferrugineum*.

  Basic signs:

  1. Winter hardiness—9 points (high).
  2. Characteristics of the plants by development phases (points):
     a. before leaving for winter—9 points;

b. after wintering—9 points;

c. before harvesting—9 points.

3. The height of the plants before harvesting—95 cm.

4. Resistance to lodging—9 points.

5. The mass of 1000 grains—49.7 g.

6. The defeat of powdery mildew and brown rust—high.

7. Grain mass from 1 m^2 to 640 g/m^2.

8. Grain mass from 1 m^2 in comparison with the standard—133.3%.

9. Grain yield from 1 hectare to 60 c/ha.

- к-54131, Sv 71536, Sweden, variety – *lutescens*.

  Basic signs:

1. Winter hardiness—9 points (high).

2. Characteristics of the plants by development phases (points):

   a. before leaving for winter—9 points;

   b. after wintering—9 points;

   c. before harvesting—9 points.

3. The height of the plants before harvesting—85 cm.

4. Resistance to lodging—9 points.

5. The mass of 1000 grains—38.7 g.

6. The defeat of powdery mildew and brown rust—high.

7. Grain mass from 1 m^2 to 630 g/m^2.

8. Grain mass from 1 m^2 in comparison with the standard—130.1%.

9. Grain yield from 1 hectare to 63 c/ha.

- к-55971, L-1749, Kursk region, variety – *lutescens*.

  Basic signs:

1. Winter hardiness—9 points (high).
2. Characteristics of the plants by development phases (points):
    a. before leaving for winter—9 points;
    b. after wintering—9 points;
    c. before harvesting—9 points.
3. The height of the plants before harvesting—100 cm.
4. Resistance to lodging—9 points.
5. The mass of 1000 grains—38.7 g.
6. The defeat of powdery mildew and brown rust—high.
7. Grain mass from 1 m^2 to 620 g/m^2.
8. Grain mass from 1 m^2 in comparison with the standard—129%.
9. Grain yield from 1 hectare to 62 c/ha.

- к-57222, Severin, Germany, variety – *lutescens*.
  Basic signs:

1. Winter hardiness—9 points (high).
2. Characteristics of the plants by development phases (points):
    a. before leaving for winter—9 points;
    b. after wintering—9 points;
    c. before harvesting—9 points.
3. The height of the plants before harvesting—95 cm.
4. Resistance to lodging—9 points.
5. The mass of 1000 grains—43.3 g.
6. The defeat of powdery mildew and brown rust—high.
7. Grain mass from 1 m^2 to 620 g/m^2.
8. Grain mass from 1 m^2 in comparison with the standard—129%.

9. Grain yield from 1 hectare to 62 c/ha.

- к-54635, Lutescens 181/75, Moscow region, variety – *lutescens*.

  Basic signs:

  1. Winter hardiness—9 points (high).
  2. Characteristics of the plants by development phases (points):
     a. before leaving for winter—9 points;
     b. after wintering—9 points;
     c. before harvesting—9 points.
  3. The height of the plants before harvesting—105 cm.
  4. Resistance to lodging—9 points.
  5. The mass of 1000 grains—46.9 g.
  6. The defeat of powdery mildew and brown rust—high.
  7. Grain mass from 1 m^2 to 615 g/m^2.
  8. Grain mass from 1 m^2 in comparison with the standard—128.1%.
  9. Grain yield from 1 hectare to 61.5 c/ha.

- к-54689, Lutescens 12424/74, Moscow region, variety – *lutescens*.

  Basic signs:

  1. Winter hardiness—9 points (high).
  2. Characteristics of the plants by development phases (points):
     a. before leaving for winter—9 points;
     b. after wintering—9 points;
     c. before harvesting—9 points.
  3. The height of the plants before harvesting—100 cm.
  4. Resistance to lodging—9 points.
  5. The mass of 1000 grains—45.5 g.
  6. The defeat of powdery mildew and brown rust—high.

7. Grain mass from 1 m^2 to 615 g/m^2.

8. Grain mass from 1 m^2 in comparison with the standard—128.1%.

9. Grain yield from 1 hectare to 61.5 c/ha.

- к-64054, Juma, Poland, variety – *lutescens*.

  Basic signs:

1. Winter hardiness—9 points (high).

2. Characteristics of the plants by development phases (points):

   a. before leaving for winter—9 points;

   b. after wintering—9 points;

   c. before harvesting—9 points.

3. The height of the plants before harvesting—100 cm.

4. Resistance to lodging—9 points.

5. The mass of 1000 grains—45.5 g.

6. The defeat of powdery mildew and brown rust—high.

7. Grain mass from 1 m^2 to 615 g/m^2.

8. Grain mass from 1 m^2 in comparison with the standard—128.1%.

9. Grain yield from 1 hectare to 61.5 c/ha.

- к-64055, Rada, Poland, variety – *lutescens*.

  Basic signs:

1. Winter hardiness—9 points (high).

2. Characteristics of the plants by development phases (points):

   a. before leaving for winter—9 points;

   b. after wintering—9 points;

   c. before harvesting—9 points.

3. The height of the plants before harvesting—80 cm.

4. Resistance to lodging—9 points.

5. The mass of 1000 grains—45.5 g.
6. The defeat of powdery mildew and brown rust—high.
7. Grain mass from 1 m^2 to 615 g/m^2.
8. Grain mass from 1 m^2 in comparison with the standard—128.1%.
9. Grain yield from 1 hectare to 61.5 c/ha.

- к-55246, Sture, Sweden, variety – *lutescens*.

  Basic signs:

  1. Winter hardiness—9 points (high).
  2. Characteristics of the plants by development phases (points):
     a. before leaving for winter—9 points;
     b. after wintering—9 points;
     c. before harvesting—9 points.
  3. The height of the plants before harvesting—90 cm.
  4. Resistance to lodging—9 points.
  5. The mass of 1000 grains—36.1 g.
  6. The defeat of powdery mildew and brown rust—high.
  7. Grain mass from 1 m^2 to 600 g/m^2.
  8. Grain mass from 1 m^2 in comparison with the standard—125%.
  9. Grain yield from 1 hectare to 60 c/ha.

- к-55315, WW 71919, Sweden, variety – *lutescens*.

  Basic signs:

  1. Winter hardiness—9 points (high).
  2. Characteristics of the plants by development phases (points):
     a. before leaving for winter—9 points;
     b. after wintering—9 points;
     c. before harvesting—9 points.

3. The height of the plants before harvesting—85 cm.
4. Resistance to lodging—9 points.
5. The mass of 1000 grains—37.7 g.
6. The defeat of powdery mildew and brown rust—high.
7. Grain mass from 1 m^2 to 600 g/m^2.
8. Grain mass from 1 m^2 in comparison with the standard—125%.
9. Grain yield from 1 hectare to 60 c/ha.

- к-55801, Lutescens 12, Kursk region, variety – *lutescens*.

  Basic signs:

  1. Winter hardiness—9 points (high).
  2. Characteristics of the plants by development phases (points):
     a. before leaving for winter—9 points;
     b. after wintering—9 points;
     c. before harvesting—9 points.
  3. The height of the plants before harvesting—75 cm.
  4. Resistance to lodging—9 points.
  5. The mass of 1000 grains—38.8 g.
  6. The defeat of powdery mildew and brown rust—high.
  7. Grain mass from 1 m^2 to 600 g/m^2.
  8. Grain mass from 1 m^2 in comparison with the standard—125%.
  9. Grain yield from 1 hectare to 60 c/ha.

- к-64053, Maltanka, Poland, variety – *lutescens*.

  Basic signs:

  1. Winter hardiness—9 points (high).
  2. Characteristics of the plants by development phases (points):
     a. before leaving for winter—9 points;

b. after wintering—9 points;

c. before harvesting—9 points.

3. The height of the plants before harvesting—75 cm.

4. Resistance to lodging—9 points.

5. The mass of 1000 grains—43.4 g.

6. The defeat of powdery mildew and brown rust—high.

7. Grain mass from 1 m^2 to 590 g/m^2.

8. Grain mass from 1 m^2 in comparison with the standard—122.9%.

9. Grain yield from 1 hectare to 59 c/ha.

- к-54657, Erythrospermum 543/75, Moscow region, variety – *erytrospermum*.

  Basic signs:

  1. Winter hardiness—9 points (high).

  2. Characteristics of the plants by development phases (points):

     a. before leaving for winter—9 points;

     b. after wintering—9 points;

     c. before harvesting—7 points.

  3. The height of the plants before harvesting—95 cm.

  4. Resistance to lodging—7 points.

  5. The mass of 1000 grains—42.6 g.

  6. The defeat of powdery mildew and brown rust—high.

  7. Grain mass from 1 m^2 to 590 g/m^2.

  8. Grain mass from 1 m^2 in comparison with the standard—122.9%.

  9. Grain yield from 1 hectare to 59 c/ha.

- к-58831, Lutescens 398, Voronezh region, variety – *lutescens*.

  Basic signs:

  1. Winter hardiness—9 points (high).

2. Characteristics of the plants by development phases (points):

    a. before leaving for winter—9 points;

    b. after wintering—9 points;

    c. before harvesting—7 points.

3. The height of the plants before harvesting—95 cm.

4. Resistance to lodging—7 points.

5. The mass of 1000 grains—46.8 g.

6. The defeat of powdery mildew and brown rust—high.

7. Grain mass from 1 m^2- 585 g/m^2.

8. Grain mass from 1 m^2 in comparison with the standard—121.8%.

9. Grain yield from 1 hectare to 58.5 c/ha.

- к-55233, Maris Marksman, England, variety – *lutescens*.

  Basic signs:

1. Winter hardiness—9 points (high).

2. Characteristics of the plants by development phases (points):

    a. before leaving for winter—9 points;

    b. after wintering—9 points;

    c. before harvesting—9 points.

3. The height of the plants before harvesting—90 cm.

4. Resistance to lodging—9 points.

5. The mass of 1000 grains—34.8 g.

6. The defeat of powdery mildew and brown rust—high.

7. Grain mass from 1 m^2 to 580 g/m^2.

8. Grain mass from 1 m^2 in comparison with the standard—120.1%.

9. Grain yield from 1 hectare to 58 c/ha.

- к-58188, FAW 34727/75, Germany, variety – *lutescens*.

  Basic signs:

  1. Winter hardiness—9 points (high).
  2. Characteristics of the plants by development phases (points):
     a. before leaving for winter—9 points;
     b. after wintering—9 points;
     c. before harvesting—9 points.
  3. The height of the plants before harvesting—90 cm.
  4. Resistance to lodging—9 points.
  5. The mass of 1000 grains—39.2 g.
  6. The defeat of powdery mildew and brown rust—high.
  7. Grain mass from 1 m^2 to 575 g/m^2.
  8. Grain mass from 1 m^2 in comparison with the standard—119.9%.
  9. Grain yield from 1 hectare—57.5 c/ha.

- к-54129, Skjaldar, Norway, variety – *lutescens*.

  Basic signs:

  1. Winter hardiness—9 points (high).
  2. Characteristics of the plants by development phases (points):
     a. before leaving for winter—9 points;
     b. after wintering—9 points;
     c. before harvesting—9 points.
  3. The height of the plants before harvesting—90 cm.
  4. Resistance to lodging—9 points.
  5. The mass of 1000 grains—39.2 g.
  6. The defeat of powdery mildew and brown rust—high.
  7. Grain mass from 1 m^2 to 575 g/m^2.

8. Grain mass from 1 m^2 in comparison with the standard—119.9%.

9. Grain yield from 1 hectare to 57.5 c/ha.

- к-57580, Liwilla, Poland, variety – *lutescens*.

  Basic signs:

1. Winter hardiness—9 points (high).

2. Characteristics of the plants by development phases (points):

   a. before leaving for winter—9 points;

   b. after wintering—9 points;

   c. before harvesting—7+ points.

3. The height of the plants before harvesting—95 cm.

4. Resistance to lodging—9 points.

5. The mass of 1000 grains—39.6 g.

6. The defeat of powdery mildew and brown rust—high.

7. Grain mass from 1 m^2- 575 g/m^2.

8. Grain mass from 1 m^2 in comparison with the standard—119.9%.

9. Grain yield from 1 hectare to 57.5 c/ha.

- к-54668, Lutescens 444/73, Moscow region, variety – *lutescens*.

  Basic signs:

1. Winter hardiness—9 points (high).

2. Characteristics of the plants by development phases (points):

   a. before leaving for winter—9 points;

   b. after wintering—9 points;

   c. before harvesting—7 points.

3. The height of the plants before harvesting—100 cm.

4. Resistance to lodging—7 points.

5. The mass of 1000 grains—40.4 g.

6. The defeat of powdery mildew and brown rust—high.

7. Grain mass from 1 m^2 to 570 g/m^2.

8. Grain mass from 1 m^2 in comparison with the standard—118.7%.

9. Grain yield from 1 hectare to 57 c/ha.

- к-64059, Roti, Germany, variety – *lutescens*.

  Basic signs:

  1. Winter hardiness—9 points (high).
  2. Characteristics of the plants by development phases (points):
     a. before leaving for winter—9 points;
     b. after wintering—9 points;
     c. before harvesting—7+ points.
  3. The height of the plants before harvesting—75 cm.
  4. Resistance to lodging—7 points.
  5. The mass of 1000 grains—42.8 g.
  6. The defeat of powdery mildew and brown rust—high.
  7. Grain mass from 1 m^2 to 565 g/m^2.
  8. Grain mass from 1 m^2 in comparison with the standard—117.7%.
  9. Grain yield from 1 hectare to 56.5 c/ha.

- к-64063, Orbis, Germany, variety – *lutescens*.

  Basic signs:

  1. Winter hardiness—9 points (high).
  2. Characteristics of the plants by development phases (points):
     a. before leaving for winter—9 points;
     b. after wintering—9 points;
     c. before harvesting—7+ points.
  3. The height of the plants before harvesting—90 cm.

4. Resistance to lodging—7 points.

5. The mass of 1000 grains—44.2 g.

6. The defeat of powdery mildew and brown rust—high.

7. Grain mass from 1 m^2 to 565 g/m^2.

8. Grain mass from 1 m^2 in comparison with the standard—117.7%.

9. Grain yield from 1 hectare to 57 c/ha.

- к-55337, Carstacht, Germany, variety – *lutescens*.

  Basic signs:

  1. Winter hardiness—9 points (high).

  2. Characteristics of the plants by development phases (points):

     a. before leaving for winter—9 points;

     b. after wintering—9 points;

     c. before harvesting—9 points.

  3. The height of the plants before harvesting—95 cm.

  4. Resistance to lodging—9 points.

  5. The mass of 1000 grains—44.2 g.

  6. The defeat of powdery mildew and brown rust—high.

  7. Grain mass from 1 m^2 to 565 g/m^2.

  8. Grain mass from 1 m^2 in comparison with the standard—117.7%.

  9. Grain yield from 1 hectare—57 c/ha.

- к-54705, Maris Kinsman, England, variety – *lutescens*.

  Basic signs:

  1. Winter hardiness—7 points (high).

  2. Characteristics of the plants by development phases (points):

     a. before leaving for winter—9 points;

     b. after wintering—9 points;

c. before harvesting—7+ points.

3. The height of the plants before harvesting—110 cm.

4. Resistance to lodging—9 points.

5. The mass of 1000 grains—46.8 g.

6. The defeat of powdery mildew and brown rust—high.

7. Grain mass from 1 m^2 to 555 g/m^2.

8. Grain mass from 1 m^2 in comparison with the standard—115.6%.

9. Grain yield from 1 hectare to 55.5 c/ha.

- к-55322, WW 72074, Sweden, a variety – *lutescens*.

Basic signs:

1. Winter hardiness—7+ points (high).

2. Characteristics of the plants by development phases (points):

a. before leaving for winter—9 points;

b. after wintering—7+ points;

c. before harvesting—7 points.

3. The height of the plants before harvesting—110 cm.

4. Resistance to lodging—9 points.

5. The mass of 1000 grains—39.2 g.

6. The defeat of powdery mildew and brown rust—high.

7. Grain mass from 1 m^2 to 550 g/m^2.

8. Grain mass from 1 m^2 in comparison with the standard—114.6%.

9. Grain yield from 1 hectare -55 c/ha.

- к-56289, Hvede Sarah, Denmark, variety – *lutescens*.

Basic signs:

1. Winter hardiness—7+ points (high).

2. Characteristics of the plants by development phases (points):

a. before leaving for winter—9 points;

b. after wintering—7 points;

c. before harvesting—7 points.

3. The height of the plants before harvesting—95 cm.

4. Resistance to lodging—9 points.

5. The mass of 1000 grains—40.6 g.

6. The defeat of powdery mildew and brown rust—high.

7. Grain mass from 1 m^2 to 550 g/m^2.

8. Grain mass from 1 m^2 in comparison with the standard—114.6%.

9. Grain yield from 1 hectare to 55 c/ha.

- к-56872, Helge, Sweden, variety – *lutescens*.

  Basic signs:

1. Winter hardiness—9 points (high).

2. Characteristics of the plants by development phases (points):

   a. before leaving for winter—9 points;

   b. after wintering—9 points;

   c. before harvesting—9 points.

3. The height of the plants before harvesting—100 cm.

4. Resistance to lodging—9 points.

5. The mass of 1000 grains—38.7 g.

6. The defeat of powdery mildew and brown rust—high.

7. Grain mass from 1 m^2 to 550 g/m^2.

8. Grain mass from 1 m^2 in comparison with the standard—114.6%.

9. Grain yield from 1 hectare to 55 c/ha.

- к-57235, Donata, Netherlands, variety – *lutescens*.

  Basic signs:

1. Winter hardiness—9 points (high).
2. Characteristics of the plants by development phases (points):
    a. before leaving for winter—9 points;
    b. after wintering—9 points;
    c. before harvesting—9 points.
3. The height of the plants before harvesting—65 cm.
4. Resistance to lodging—9 points.
5. The mass of 1000 grains—38.1 g.
6. The defeat of powdery mildew and brown rust—high.
7. Grain mass from 1 m^2 to 550 g/m^2.
8. Grain mass from 1 m^2 in comparison with the standard—114.6%.
9. Grain yield from 1 hectare to 55 c/ha.

- к-64027, Bussard, Germany, variety – *lutescens*.

  Basic signs:

1. Winter hardiness—7 points (high).
2. Characteristics of the plants by development phases (points):
    a. before leaving for winter—9 points;
    b. after wintering—9 points;
    c. before harvesting—7 points.
3. The height of the plants before harvesting—100 cm.
4. Resistance to lodging—9 points.
5. The mass of 1000 grains—40.0 g.
6. The defeat of powdery mildew and brown rust—high.
7. Grain mass from 1 m^2 to 545 g/m^2.
8. Grain mass from 1 m^2 in comparison with the standard—113.5%.

9. Grain yield from 1 hectare to 54.5 c/ha.

- κ-64028, Faktor, Germany, variety – *lutescens*.

  Basic signs:

  1. Winter hardiness—7 points (high).
  2. Characteristics of the plants by development phases (points):
     a. before leaving for winter—9 points;
     b. after wintering—9 points;
     c. before harvesting—7+ points.
  3. The height of the plants before harvesting—85 cm.
  4. Resistance to lodging—9 points.
  5. The mass of 1000 grains—42.1 g.
  6. The defeat of powdery mildew and brown rust—high.
  7. Grain mass from 1 m^2 to 545 g/m^2.
  8. Grain mass from 1 m^2 in comparison with the standard—113.5%.
  9. Grain yield from 1 hectare to 54.5 c/ha.

- κ-55313, WW 71822, Sweden, variety – *lutescens*.

  Basic signs:

  1. Winter hardiness—9 points (high).
  2. Characteristics of the plants by development phases (points):
     1. before leaving for winter—9 points;
     2. after wintering—9 points;
     3. before harvesting—9 points.
  3. The height of the plants before harvesting—85 cm.
  4. Resistance to lodging—9 points.
  5. The mass of 1000 grains—35.2 g.
  6. The defeat of powdery mildew and brown rust—high.

7. Grain mass from 1 m^2 to 540 g/m^2.

8. Grain mass from 1 m^2 in comparison with the standard—112.5%.

9. Grain yield from 1 hectare to 54 c/ha.

- к-64025, Muck, Germany, variety – *lutescens*.

  Basic signs:

  1. Winter hardiness—7 points (high).
  2. Characteristics of the plants by development phases (points):
     a. before leaving for winter—9 points;
     b. after wintering—9 points;
     c. before harvesting—7 points.
  3. The height of the plants before harvesting—100 cm.
  4. Resistance to lodging—9 points.
  5. The mass of 1000 grains—40.4 g.
  6. The defeat of powdery mildew and brown rust—high.
  7. Grain mass from 1 m^2 to 535 g/m^2.
  8. Grain mass from 1 m^2 in comparison with the standard—111.5%.
  9. Grain yield from 1 hectare to 53.5 c/ha.

- к-54823, Linos, Germany, variety – *lutescens*.

  Basic signs:

  a. Winter hardiness—7 points (high).

  b. Characteristics of the plants by development phases (points):
     a. before leaving for winter—9 points;
     b. after wintering—9 points;
     c. before harvesting—7 points.

  c. The height of the plants before harvesting—105 cm.

  d. Resistance to lodging—7+ points.

e. The mass of 1000 grains—43.7 g.

f. The defeat of powdery mildew and brown rust—high.

g. Grain mass from 1 m^2 to 535 g/m^2.

h. Grain mass from 1 m^2 in comparison with the standard—111.5%.

i. Grain yield from 1 hectare to 53.5 c/ha.

- к-55218, C 975/69, Poland, variety – *lutescens*.

  Basic signs:

  1. Winter hardiness—7 points (high).
  2. Characteristics of the plants by development phases (points):
     a. before leaving for winter—9 points;
     b. after wintering—9 points;
     c. before harvesting—7 points.
  3. The height of the plants before harvesting—90 cm.
  4. Resistance to lodging—9 points.
  5. The mass of 1000 grains—42.6 g.
  6. The defeat of powdery mildew and brown rust—high.
  7. Grain mass from 1 m^2 to 535 g/m^2.
  8. Grain mass from 1 m^2 in comparison with the standard—111.5%.
  9. Grain yield from 1 hectare to 53.5 c/ha.

- к-54563, Lutescens 755/76, Moscow region, variety – *lutescens*.

  Basic signs:

  1. Winter hardiness—9 points (high).
  2. Characteristics of the plants by development phases (points):
     a. before leaving for winter—9 points;
     b. after wintering—9 points;
     c. before harvesting—9 points.

3. The height of the plants before harvesting—95 cm.
4. Resistance to lodging—7+ points.
5. The mass of 1000 grains—35.8 g.
6. The defeat of powdery mildew and brown rust—high.
7. Grain mass from 1 m^2 to 535 g/m^2.
8. Grain mass from 1 m^2 in comparison with the standard—111.4%.
9. Grain yield from 1 hectare to 53.5 c/ha.

- к-55321, WW 72073, Sweden, variety – *lutescens*.

  Basic signs:

  1. Winter hardiness—7+ points (high).
  2. Characteristics of the plants by development phases (points):
     a. before leaving for winter—9 points;
     b. after wintering—9 points;
     c. before harvesting—7 points.
  3. The height of the plants before harvesting—85 cm.
  4. Resistance to lodging—9 points.
  5. The mass of 1000 grains—39.4 g.
  6. The defeat of powdery mildew and brown rust—high.
  7. Grain mass from 1 m^2 to 530 g/m^2.
  8. Grain mass from 1 m^2 in comparison with the standard—110.0%.
  9. Grain yield from 1 hectare to 53 c/ha.

- к-55306, Sv. 01750, Sweden, variety – *lutescens*.

  Basic signs:

  1. Winter hardiness—9 points (high).
  2. Characteristics of the plants by development phases (points):
     a. before leaving for winter—9 points;

DOI: http://dx.doi.org/10.5772/intechopen.109293

   b. after wintering—9 points;

   c. before harvesting—7 points.

3. The height of the plants before harvesting—85 cm.

4. Resistance to lodging—9 points.

5. The mass of 1000 grains—39.5 g.

6. The defeat of powdery mildew and brown rust—high.

7. Grain mass from 1 m^2 to 530 g/m^2.

8. Grain mass from 1 m^2 in comparison with the standard—110.0%.

9. Grain yield from 1 hectare to 53 c/ha.

The yield from the above samples is compared to the standard cultivar Moscow 39, the yield of which was 48.0 c/ha (on average out of 14 plots with area of 2 m^2).

3. Conclusions

Any breeding program, including the creation of drought-resistant varieties, requires a rich and well-researched material, which is concentrated within the VIR World Collection. The success of the work can be ensured with a large volume of hybridization work and a wide selection of drought-resistant forms; therefore, diverse genetic material should be involved for crosses and targeted selection for homeostasis should be carried out, against which the genes of drought resistance, yield, mass of 1000 grains, resistance to diseases will be combined. In our studies on drought resistance, German and Russian wheat had the highest percentage of resistant forms.

Acknowledgements

The presented study was carried out as part of the Governmental Task for the Federal Horticultural Center for Breeding, Agrotechnology and Nursery # 0432-2021-0003 to preserve, enlarge, and study the genetic collections of agricultural plants and establish the repositories of the horticultural and small-fruit crops that are free from hazardous viruses; and as a part of the Subject Plan for the All-Russian Research Institute of Phytopathology # 0598-2019-0005, Official Registration Number EGISU R&D–AAAA-A19-1191212901090. This paper has been supported by the Kazan Federal University Strategic Academic Leadership Program (PRIORITY-2030).

Author details

Sulukhan Temirbekova[1*], Ivan Kulikov[2], Natalya Ionova[3], Yuliya Afanasyeva[2] and Elena Kalashnikova[4]

1 All-Russian Research Institute of Phytopathology, Bolshye Vyazyomy, Odintsovo District, Moscow, Russia

2 Federal Horticultural Center for Breeding, Agrotechnology and Nursery, Moscow, Russia

3 Institute of Fundamental Medicine and Biology, Kazan Federal University, Kazan, Russia

4 Russian State Agrarian University, Moscow Agricultural Academy named after K.A. Timiryazev, Moscow, Russia

*Address all correspondence to: sul20@yandex.ru

References

[1] Budyko MI. Evolution of the Biosphere. Leningrad: Hydrometeoizdat; 1984. p. 488

[2] Vavilov NI. World resources of drought-resistant cultivars. In: Reports of the All-Union Conference on Combating Drought. Vol. 2. Leningrad: VIR; 1931. pp. 18-28

[3] Krupnov VA. Drought and wheat breeding: A systematic approach. Agricultural Biology. 2011;**1**: 12-23

[4] Passioura JB. The drought environment: Physical, biological and agricultural perspectives. Journal of Experimental Botany. 2007;**58**:113-117. DOI: 10.1093/jxb/erl212

[5] Chen J, Lin L, Lu G. An index of soil drought intensity and degree: An application on corn and a comparison with CWSI. Agricultural Water Management. 2010;**97**:865-871. DOI: 10.1016/J.AGWAT.2010.01.017

[6] Guttieri MJ, Stark JC, O'Brien K., Souza E. Relative sensitivity of spring wheat grain yield and quality parameters to moisture deficit. Crop Science. 2001; **41**:327-335. DOI: 10.2135/ CROPSCI2001.412327X

[7] Krupnov VA, Germantsev LA, Krupnova OV. The effect of a temperature increase on the grain weight of the spring wheat in the Volga regions. Annual Wheat Newsletter (USA). 2001; **47**:145

[8] Blumenthal CS, Barlow EWR, Wrigley CW. Growth environment and wheat quality: The effect of heat stress on dough properties and gluten proteins. Journal of Cereal Science. 1993;**18**:3-21. DOI: 10.1006/JCRS.1993.1030

[9] Triboi E, Martre P, Triboi-Blondel AM. Environmentally-induced changes in protein composition in developing grains of wheat are related to changes in total protein content. Journal of Experimental Botany. 2003;**54**: 1731-1742. DOI: 10.1093/JXB/ERG183

[10] Irmac S, Naaem HA, Lookhart GL, Mac RF. Effect of heat stress on wheat proteins during kernel development in wheat nearisogenic lines differing at Glu- D1. Journal Cereal Science. 2008; **48**:513-516

[11] Temirbekova SK, Cheremisova TD, Kulikov IM, Afanasyeva YV, Zuev EV, Potapova ES. The Source Material of Winter Wheat for the "Ideal Cultivar" According to N.I. Vavilov for Resistance to Abiotic and Biotic Stress Factors of the Central Region of the Russian Federation. Moscow: VSTISP; 2020. p. 108